CONCISE A-LEVEL
PHYSICS SI EDITION

CONCISE A-LEVEL PHYSICS SI EDITION

A. R. W. HAYES

D.F.C., B.Sc.(Lond.), A.R.C.O.

*Senior Science Master, Queen
Elizabeth's School, Barnet*

LONGMAN

Longman
1724-1974

LONGMAN GROUP LIMITED
London

Associated companies, branches and representatives
throughout the world

© A. R. W. Hayes 1962
© Longman Group Limited 1974

First published 1962
Second edition 1969
Third edition 1974

ISBN 0 582 35133 2

Printed in Great Britain by
Fletcher & Son Ltd, Norwich

Preface to SI Edition

This edition brings the text fully into line with the recommendations of the A.S.E. Report *SI Units, Signs, Symbols, and Abbreviations*.

The changes in detail are extensive, and certain sections are re-drafted. In particular, the Electromagetism and Electrostatics sections now follow the simplified '*B* and *E* vectors' treatment, with the inverse square law approach in Electrostatics. It is hoped that this treatment will in future prove acceptable to most schools.

1974   A.R.W.H.

Preface to the First Edition

These notes, based on duplicated revision sheets which the writer has used for several years with his own Sixth Forms, are compiled with two principal objectives in view: to reduce the need for note-taking in class, and to provide a basis for a long-term revision programme which can be carried out by the student with a minimum of assistance.

Many physics candidates show, by their lack of understanding of basic principles, the need for a more thoughtful and systematic revision of basic work. This can hardly be achieved in a scramble to 'get by' confined to the last few weeks preceding the public examinations. A more extended programme, however, is not likely to succeed without careful guidance at every stage. But the teacher has little class time available for constantly reviewing earlier work. The limitations of full-scale textbooks as revision aids are well known: at this stage a more concise version is needed in which essential points stand out more clearly. Class written notes take up a great deal of time. Though some may be necessary, the mechanical process of copying from dictation or the blackboard is in itself of little value as an aid to memory, and indeed can prevent the pupil taking a more active part in classwork. Some teachers prefer their pupils to make their own notes; while these may be of greater value in the making, they need careful checking if they are to be used as the only revision aid.

The present notes are an attempt to meet these difficulties in a practical way. They comprise a reasonably comprehensive summary of basic Advanced Level topics, in a form suitable both for the quick 'last-minute' revision and also for the more detailed revision which should precede it. Though not intended to replace normal textbooks, they include all the logical argument essential to a real understanding of the work—anything less would merely encourage the memorizing of unconnected facts, which is contrary to the whole spirit and purpose of a physics course. 'O' level fundamentals are included where necessary. The grouping and sequence of material has received particular attention. The student following a normal physics course at 'A' level or its equivalent should be able to work through the notes with little recourse to any outside help; they may thus, it is hoped, facilitate the planning by the teacher of a long-term revision programme (extending over perhaps the whole of the second year of a two-year course). Such a programme must demand a minimum of supervision and direction if it is to be practicable. As class notes can be greatly reduced, the saving of class time makes possible in the second year a 'weekly revision test', which is both valuable to the pupil and essential to the working of such a scheme.

Present Advanced Level requirements are closely followed, but the aim is not, of course, an exhaustive treatment of every possible topic: particular needs can be met by supplementary notes. In Light, both Real-is-Positive and New Cartesian sign conventions are worked, on an equal basis. The lens formula is derived by refraction at two spherical surfaces and also by the deviation method as an alternative.

A point is made of careful definition of quantities and clear statements of laws, and a special Index to these is provided for easy reference. It is hoped that a suitable balance between the needs of candidates of varying mathematical attainments has been achieved; calculus *notation* is freely used, but only the simplest differentiations and integrations are called for, and these can be otherwise dealt with if required. A few worked examples are included as an integral part of the text, but the provision of numerical problems is outside the scope of this book.

I am greatly indebted to the publishers and their reader, and to my colleague Mr D. W. Fairbairn, B.Sc., for much help and advice at all stages during the preparation of the work.

October 1961   A.R.W.H.

Introduction

Revision. Understanding is of prime importance. It is an insult to the intelligence to learn parrot-fashion work which is not understood. It is also bad policy, because facts 'learnt' in this way are soon forgotten, whereas facts that are reasoned are retained almost indefinitely. There is no danger of work being wasted by starting revision too early, provided it is done properly. The bulk of revision should be spread over a long period, with a final refresher in the last few weeks.

When revising, work through a section slowly and critically, making sure each step in the argument is correct. Then try *writing it out* from memory. Few students are prepared to do this at first, but it is the only sure way of finding out weaknesses, and of developing the speed necessary in examinations. It cannot be too strongly emphasized that just looking through notes, without the use of pencil and paper, is no insurance that they have been grasped.

To sum up:

> Start revising early,
> Understand each step,
> Write out from memory.

Preparing for examinations. One problem in examinations is the time factor. Much can be done to prepare for this beforehand. Firstly, by writing out standard bookwork from memory, so as to have it at the fingertips. Standard work should be known so well that it can be disposed of quickly, leaving more time for calculations and unfamiliar parts of questions.

Secondly, by practising setting down information in the most compact form with the least waste of words—including all the necessary facts but without repetition. A number of practice answers should be written out in full, keeping strictly to the time allocation.

In the examination. There is no substitute for knowing the work, but many candidates with the knowledge that they do possess could gain substantially higher marks by observing a few simple rules in the examination room:

Answer precisely the question asked. This is likely to be based on work that has been covered, but the question may be in an unfamiliar form, and the answer must be adapted accordingly. Answer each part of the question separately. Make a special note of the phrasing used, particularly of such words as 'define', 'explain', 'state', 'derive', 'describe'—these words are meant to be taken quite literally, and each has a different meaning.

Attempt the full number of questions specified, even if you cannot answer them all fully. The last few marks in one question are generally more difficult to obtain than the first few in another, and the candidate who answers, say, four questions instead of five reduces his maximum to 80% with probably little corresponding advantage.

Begin by answering the questions you can do best, but do not spend more than a small overtime on any one question. In a descriptive question, the best guide to the amount of detail expected is the time available for answering it. Try not to be caught by the clock at the end of the examination just about to earn some easy marks on a straightforward part—this is bad timing.

Derivations. 1. When writing these out, first explain briefly what you are doing, and define your symbols. Formulae stated without mention of what the symbols stand for are meaningless.

2. State clearly any approximations used, and on what grounds.

3. Give brief reasons for each step.

Descriptions of experiments. 1. A labelled diagram saves part of the verbal description—information shown on a diagram need not be repeated in the account. Diagrams should be clear and neat, but do not spend a disproportionate length of time drawing them.

2. Give complete down-to-earth details of the procedure adopted, from start to finish. State what measurements have to be taken. Avoid the expression 'the result may be calculated'—explain exactly *how* it is worked out from the observations made. This often involves quoting a formula; unless asked for, a theoretical proof of the formula is not required.

3. Mention the most important precautions necessary to ensure accuracy.

Laws, definitions and formulae. There is much to be said for learning the wording of the more difficult laws and definitions more or less by heart: precise wording is very important. Formal statements of laws and definitions are normally indicated in these notes by smaller type, with the quantity being defined, or the name of the law, in SMALL CAPITALS.

Those formulae which are displayed *and numbered* can be regarded as key formulae to be remembered.

Outside reading and numerical examples. These notes cover basic work only, and are not intended to limit outside reading—rather, by saving wasted effort, should they leave more time for it.

Finally, it is only by practice with every type of descriptive and numerical question that facility comes. Very few numerical examples are included in these notes.

The Greek Alphabet

Capitals in brackets

Alpha, a	Iota, ι	Rho, ρ
Beta, β	Kappa, κ	Sigma, σ $[\Sigma]$
Gamma, γ	Lambda, λ	Tau, τ
Delta, δ $[\Delta]$	Mu, μ	Upsilon, υ
Epsilon, ε	Nu, ν	Phi, ϕ $[\Phi]$
Zeta, ζ	Xi, ξ	Chi, χ
Eta, η	Omikron, o	Psi, ψ $[\Psi]$
Theta, θ	Pi, π	Omega, ω $[\Omega]$

Contents

IV—SOUND

V—ELECTRICITY

VI—MODERN PHYSICS

Note on SI Units

The basic SI units are the metre, kilogramme, second, ampere, kelvin, candela and mole. Certain derived units also have special names (see list below). Other derived units should be expressed in index notation (e.g. Pressure in $N\ m^{-2}$), not in the solidus (oblique stroke) form. Note that the letter s is never added to symbols for units in the plural.

All derived SI units are 'coherent', i.e. are obtained from the basic units by multiplication, without introducing numerical factors. Multiples and sub-multiples of the units may be introduced, but those involving powers which are multiples of three are preferred e.g. 10^3, 10^6, 10^{-3}, etc.

$\times 10^3$	kilo-	k	$\times 10^{-3}$	milli-	m
$\times 10^6$	mega-	M	$\times 10^{-6}$	micro-	μ
$\times 10^9$	giga-	G	$\times 10^{-9}$	nano-	n
$\times 10^{12}$	tera-	T	$\times 10^{-12}$	pico-	p

Length	metre	m
Mass	kilogramme	kg
Time	second	s
Force	newton	N
Energy or Work	joule	J
Power	watt	W
Angle	radian	rad
Frequency	hertz	Hz
Luminous intensity	candela	cd
Luminous flux	lumen	lm
Illumination	lux	lx
Heat energy	joule	J
Common temperature	degree Celsius	°C
Thermodynamic temperature	kelvin	K
Amount of a substance	mole	mol
Electric current	ampere	A
Electric charge	coulomb	C
E.m.f., P.d. or Electric potential	volt	V
Resistance	ohm	Ω
Magnetic flux	weber	Wb
Magnetic flux density	tesla	T
Inductance, mutual or self	henry	H
Capacitance	farad	F

Part I
General Physics

1. Mechanics: Laws and Definitions

Statics

A SCALAR quantity—e.g. speed, energy, density—has magnitude only.

A VECTOR quantity—e.g. velocity, acceleration, force, momentum—has both magnitude and direction.

PARALLELOGRAM OF FORCES. If two forces acting at a point are represented in magnitude and direction by the sides AB, AD of a parallelogram $ABCD$, their resultant is represented in magnitude and direction by the diagonal AC.

TRIANGLE OF FORCES. If three forces acting at a point can be represented in magnitude and direction by the three sides of a triangle, taken in order, then they are in equilibrium.

The MOMENT OF A FORCE ABOUT A POINT is the product of the force and the perpendicular distance from the point to the line of action of the force.

PRINCIPLE OF MOMENTS. If a body is in equilibrium under the action of a number of coplanar forces, the algebraic sum of the moments of the forces about any point in the plane is zero.

A COUPLE consists of two equal and opposite forces not in the same straight line.

The MOMENT OF A COUPLE or TORQUE is the product of one of the forces and the perpendicular distance between their lines of action.

The WEIGHT of a body is the gravitational force acting upon it.

The CENTRE OF GRAVITY of a body is the point through which the whole of its weight may be considered to act.

The WORK DONE by a force is equal to the product of the force and the distance it moves in the direction of its line of action.

The ENERGY of a system is its capacity for doing work, measured by the amount of work it can do.

Machines

The force applied to a machine is called the EFFORT, and the force the machine overcomes is called the LOAD.

MECHANICAL ADVANTAGE, M.A. = Load/Effort.

VELOCITY RATIO, V.R. $= \dfrac{\text{Distance moved by effort}}{\text{Corresponding distance moved by load}}$.

EFFICIENCY = Work got out/Work put in.

1 . MECHANICS: LAWS AND DEFINITIONS

It can be shown from the above that Efficiency $= \dfrac{\text{M.A.}}{\text{V.R.}}$.

Friction

LAWS OF FRICTION. (1) The limiting frictional force F is proportional to the normal reaction N between the surfaces.

(2) The limiting frictional force is independent of the area of contact of the surfaces.

(3) The kinetic or sliding frictional force is slightly less than the limiting static frictional force, but is independent of the relative velocity of the surfaces.

COEFFICIENT OF STATIC FRICTION $(\mu) = \dfrac{\text{Limiting friction } F}{\text{Normal reaction } N}$,

or
$$F = \mu N \dotfill (1.1)$$

The ANGLE OF FRICTION (λ) is the angle between N and the resultant of F and N.

Thus
$$\mu = \tan \lambda \dotfill (1.2)$$

Hydrostatics

PRESSURE $=$ Force per unit area. *Unit:* $N\,m^{-2}$.

The PRESSURE AT A POINT $= \lim\limits_{\text{area} \to 0} \left(\dfrac{\text{Force}}{\text{Area}} \right)$. The force is a push force, or thrust, and is measured perpendicular to the area.

DENSITY $=$ Mass per unit volume. *Unit:* $kg\,m^{-3}$.

ARCHIMEDES' PRINCIPLE. When a body is totally or partially immersed in a fluid it experiences an upthrust, or loss in weight, equal to the weight of fluid displaced. The upthrust acts through the centre of gravity of the displaced fluid.

LAW OF FLOTATION. When a body floats in a liquid it displaces its own weight of liquid.

BOYLE'S LAW. For a fixed mass of any gas at constant temperature, the pressure is inversely proportional to the volume.

Dynamics

NEWTON'S LAWS OF MOTION. (1) A body remains in a state of rest or uniform motion in a straight line unless acted upon by an external force.

(2) The rate of change of momentum of a body is proportional to the applied external force, and takes place in the direction in which the force acts.

(3) When two bodies A, B exert forces on each other the force (action) of A on B is always equal and opposite to the force (reaction) of B on A.

MOMENTUM. The momentum a body possesses at a given instant is defined as the product of its mass and its velocity. It is a vector.

3

GENERAL PHYSICS

The NEWTON is the unit of *force*. The newton is that force which gives to a mass of 1 kilogramme an acceleration of 1 metre per second per second.

The JOULE is the unit of *energy* or *work*. 1 joule of work is done when a force of 1 newton acts through 1 metre.

POWER is rate of doing work, or work done per unit time.

The WATT is the unit of *power*. It is a rate of working of 1 joule per second.

KINETIC ENERGY is energy possessed by a body by virtue of its motion. It may be of two sorts: translational and rotational.

POTENTIAL ENERGY is energy possessed by a body by virtue of its position. It arises from the fact that work has to be done getting the body into this position, e.g. in raising a weight, or compressing a spring.

CONSERVATION OF ENERGY. Energy cannot be created or destroyed, but can be converted from one form to another.

CONSERVATION OF MOMENTUM. When bodies in a system interact there is no resultant change of momentum in any direction, provided that no external force acts.*

The COEFFICIENT OF RESTITUTION (*e*) for two colliding bodies is the ratio $\frac{\text{Relative velocity of separation}}{\text{Relative velocity of approach}}$. For two colliding spheres, these velocities are measured along their line of centres.

The MOMENT OF INERTIA (*I*) of a rigid body about a given axis is defined as Σmr^2, where *m* is the mass of each small element of the body, and *r* is the distance of the element from the axis of rotation. *Unit:* kg m².

Concepts of force and mass. Force is a useful concept, arising from our sense of touch—we imagine we can 'feel' forces. It is interesting to contemplate whether such a concept would have arisen if we had possessed only the senses of sight, hearing and smell.

The unit of force is defined in terms of Newton's second law of motion. Does this mean that this so-called law is merely a definition of force? We could use any one of a number of the laws of mechanics to define force—Hooke's law, Boyle's law, the law of conservation of energy, the inverse square law of forces, the law of gravitation, etc. But that law would then lose its significance as a 'law' and become just a definition of force. Alternatively, the laws of mechanics could be regarded as merely a self-consistent system, in which the concept of force is the unifying element, enabling all the laws to be simple ones, and therefore plausible.

* Momentum should not be confused with kinetic energy. Momentum is indestructible. Unlike kinetic energy, it cannot be converted to some other form.

In fact, the idea of force is so ingrained that we are prepared to accept it as a reality without sacrificing any law of mechanics explicitly to define it. Although we define the *unit* of force in terms of Newton's second law, we still regard that law as something more than just a definition.

Our concept of mass, again, is easy to grasp, because it is bound up with the ideas of bulk and conservation of matter. Just as force is the external agency acting upon a body, causing it to change its motion, so mass is the intrinsic quality of the body itself which causes it to resist such changes in motion.

Mass and weight. Both quantities are linked in our minds with bulk, and we often loosely use the same unit for both. Scientifically, however, they must be carefully distinguished:

The INERTIA or MASS of a body is the property by which it tends to resist changes in motion. Numerically it is the constant m in the formula $F = ma$ (Eqn. 2.1). *Unit:* kg.

The WEIGHT of a body is the gravitational force acting upon it. *Unit:* N.

The weight W is given by $W = mg$ (Eqn. 2.3). Therefore, unlike the mass, the weight is not constant, but changes as the acceleration due to gravity g changes from place to place.

2. Mechanics: Derivations

Formulae from Newton's second law

(1) *Translational motion.* Stating the second law mathematically,

$$F \propto \frac{d}{dt}(mv), \quad \text{or} \quad F \propto m\frac{dv}{dt} \text{ (if } m \text{ is constant).}$$

But dv/dt is the acceleration a. Therefore $F \propto ma$, or $F = kma$, where k is a constant. Choosing $k = 1$ defines the newton (p. 4).

$$\therefore \qquad F = ma \quad \dots\dots\dots\dots\dots\dots\dots\dots(2.1)$$

Units: m in kg, a in m s^{-2}, F in N.

GENERAL PHYSICS

(2) *Rotational motion.*
A force *F* applied to an
element, mass *m*, of a rigid
body causes it to rotate
about a fixed axis *O*
(Fig. 1). Applying Eqn.
(2.1), $F = ma = mra$,
where *a* is the angular
acceleration about *O*.

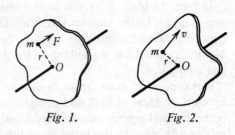

Fig. 1.   Fig. 2.

The moment of the couple, or torque, applied to the particle about
O is $Fr = mr^2a$. The total torque *T* applied to the body $= \Sigma mr^2 a$.
All the particles have the same angular acceleration *a*, and Σmr^2 is
called the moment of inertia *I* of the body about the axis *O* (p. 4).

$$\therefore \qquad\qquad T = Ia \dots\dots\dots\dots\dots\dots\dots(2.2)$$

(3) *Weight of a body.* Applying Eqn. (2.1) to a body falling freely
under gravity, where the only force acting is its own weight *W*, and
the acceleration is *g*,

$$W = mg \dots\dots\dots\dots\dots\dots\dots\dots(2.3)$$

Kinetic energy

(1) *Translational energy.* The loss in potential energy of a body,
mass *m*, falling freely through a height *h* is

$$\text{Force} \times \text{Distance} \quad \text{or} \quad mgh.$$

This energy becomes kinetic energy. Substituting in the constant
acceleration formula $v^2 = u^2 + 2as$ (p. 16) we obtain $v^2 = 2gh$.
Kinetic energy gained $= mgh = mg\,(v^2/2g) = \frac{1}{2}mv^2$.

$$\therefore \qquad\text{Translational kinetic energy} = \tfrac{1}{2}mv^2 \dots\dots\dots(2.4)$$

(2) *Rotational energy.* Consider an element, mass *m*, of a rigid
body rotating about a fixed axis *O* with velocity *v* (Fig. 2). Its dis-
tance from the axis is *r*. Its kinetic energy $= \frac{1}{2}mv^2 = \frac{1}{2}mr^2\omega^2$, where
ω is the angular velocity about *O*.

The K.E. of the whole body $= \frac{1}{2}\Sigma mr^2\omega^2$, where ω is the same for
all the elements, and Σmr^2 is the moment of inertia *I* about this axis.

$$\therefore \qquad\text{Rotational kinetic energy} = \tfrac{1}{2}I\omega^2 \dots\dots\dots(2.5)$$

Work done

(1) *Against a force.* When a force F moves through a distance s in the direction of the force,

$$\text{Work done} = Fs \dots\dots\dots\dots\dots\dots(2.6)$$

(2) *Against a torque.* When a force F acts at a distance r from the axis of rotation O, the torque exerted is $T = Fr$. The work done by the force moving a distance s round the circumference, or through an angle θ (where $s = r\theta$) is $Fs = Fr\theta = T\theta$.

$\therefore$ $$\text{Work done} = T\theta \dots\dots\dots\dots\dots\dots(2.7)$$

Uniform circular motion. A particle moves at uniform speed v round a circle of radius r (Fig. 3). In a time δt it travels through an angle $\delta\theta$ as shown, and its change in velocity is δv (Fig. 3a). As δt and $\delta\theta \to 0$, the direction of δv tends to be perpendicular to v. The instantaneous acceleration is therefore directed towards the centre of the circle.

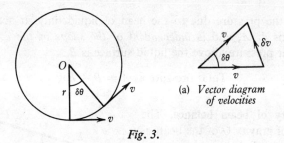

(a) *Vector diagram of velocities*

Fig. 3.

From the vector diagram, $\delta v = v\delta\theta$. Therefore

$$\text{Acceleration} = \frac{dv}{dt} = v\frac{d\theta}{dt} = v\omega = \frac{v^2}{r} = r\omega^2$$

where ω is the angular velocity of the particle.

$\therefore$ $$\text{Acceleration towards centre} = \frac{v^2}{r} \dots\dots\dots\dots(2.8)$$

For a body, mass m, travelling in a circle a force F must be continuously applied to it to maintain the acceleration towards the centre. This is the *centripetal force*, and is given by $F = ma$

or $$\text{Centripetal force} = \frac{mv^2}{r} \dots\dots\dots\dots(2.9)$$

7

The body is not in equilibrium, so there is no corresponding outwards, or *centrifugal*, force acting upon it. If we consider, for example, the motion of a stone whirled in a horizontal circle on the end of a string, we see that the only radial force acting *on the stone* is the inwards centripetal force, which is exerted by the string to maintain the acceleration towards the centre. There is of course an equal and opposite centrifugal reaction on the person at the centre holding the string, but this is not a force acting on the body being considered, i.e. the stone.

Pressure in a liquid. Consider a vertical column of liquid, of cross-sectional area A, depth h, density ρ, and mass m (Fig. 4). Force on area A at base of column = Weight of liquid vertically above = mg = $hA\rho g$. Pressure is Force/Area = $h\rho g$.

With h in m, ρ in kg m^{-3}, g in m s^{-2},

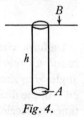

Fig. 4.

$$\text{Pressure} = h\rho g \text{ in N m}^{-2} \quad \dots\dots\dots\dots(2.10)$$

This is the pressure due to the head of liquid only. It acts in all directions at A, and is *independent of the shape of the container*. If the air pressure above the liquid surface is B,

$$\text{Total pressure at } A = B + h\rho g \quad \dots\dots\dots(2.11)$$

Theory of beam balance. The centre of gravity G of the beam is a distance h below the central knife-edge O (Fig. 5). The line joining the two scale pan knife-edges is a distance x below O. The arms of the balance are each of length L. The mass of the beam is M, and of each scale pan, m.

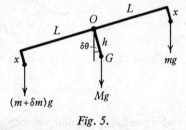

Fig. 5.

Let a mass δm added to one pan cause a deflection $\delta\theta$ in radian.

Taking L.H. moments about O, $(m + \delta m)g(L \cos \delta\theta - x \sin \delta\theta)$,

and R.H. moments, $mg(L \cos \delta\theta + x \sin \delta\theta) + Mgh \sin \delta\theta$.

As $\delta\theta$ is small, $\cos \delta\theta \simeq 1$, and $\sin \delta\theta \simeq \delta\theta$.

$$\therefore \quad mL - mx\delta\theta + \delta mL - \delta mx\delta\theta = mL + mx\delta\theta + Mh\delta\theta.$$

8

As $\delta m x \delta \theta$ is of the second order of small quantities it can be ignored, and the expression reduces to

$$\delta m L = \delta \theta (2mx + Mh).$$

Now, SENSITIVITY $= \dfrac{\text{Change in deflection produced}}{\text{Change in quantity causing deflection}},$

$\therefore$ Sensitivity $= \dfrac{\delta \theta}{\delta m} = \dfrac{L}{2mx + Mh}$ ....(2.12)

Distinguish between *sensitivity* and *accuracy*. What factors determine the sensitivity of the beam balance? What is the condition necessary for the sensitivity to be independent of the load?

Theoretical proof of Archimedes' principle. The space occupied by the immersed body X (Fig. 6) was previously occupied by the fluid, and this fluid was in equilibrium; therefore its own weight, acting downwards, must have been exactly balanced by the upthrust exerted on it by the surrounding fluid.

Fig. 6.

The same upthrust is now exerted on the immersed body, an upthrust equal to the weight of fluid displaced.

Conservation of momentum deduced from Newton's laws. By the third law of motion, when any two bodies in a system interact, the force on the one is accompanied at every instant by an equal and opposite force on the other; therefore no *net* force ever acts within a closed system. By the second law, no momentum can therefore be created within the system, if no external force acts.

3. Mechanics: Problems

This Section deals briefly with mathematical principles, and a few worked examples are given illustrating particular points. It is emphasized that a mere *knowledge* of principles is not sufficient—much practice is needed in applying them in a variety of examples.

9

GENERAL PHYSICS

Combining and resolving forces. It is often necessary to combine two coplanar forces into one force; or conversely to split up one force into two forces in given directions:—

(1) The resultant of two forces acting at a point is obtained by the parallelogram law.

(2) The resultant of two like parallel forces of magnitudes P, Q is a similar parallel force of magnitude $(P + Q)$, acting through a point B such that $P \cdot AB = Q \cdot BC$ (Fig. 7a).

(3) The resultant of two unlike parallel forces of magnitudes P, Q, where $P > Q$, is a parallel force of magnitude $(P - Q)$, acting through B such that $P \cdot AB = Q \cdot BC$ (Fig. 7b).

(4) When two unlike parallel forces are equal in magnitude they have no resultant, but form a *couple*. Their effect is entirely rotational and their turning moment is the same about all points in the plane.

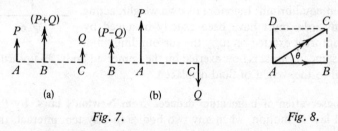

Fig. 7.   Fig. 8.

(5) To split up a given force AC into two components AB, AD in given directions, the parallelogram law is applied in reverse. If the two directions are at right angles (Fig. 8),

$$AB = AC \cos \theta \quad \text{and} \quad AD = AC \sin \theta.$$

Forces in equilibrium. For complete equilibrium of a body the forces on it must satisfy the conditions for both translational and rotational equilibrium.

(1) Method for three non-parallel forces. For translational equilibrium, the resultant of any two must be equal in magnitude, and opposite in direction, to the third; alternatively, the triangle of forces may be applied. For rotational equilibrium, the lines of action of the forces must all pass through one point.

(2) Method for any number of forces. The forces are first split up into components in two convenient directions at right angles. For translational equilibrium, the algebraic sum of the components in each of these two directions must be zero; further, for rotational

10

equilibrium, the algebraic sum of their moments about any one point must also be zero.

Forces not in equilibrium. If the forces on a body do not balance, a net force acts, and a linear acceleration occurs in the direction of this force. Likewise, a net torque produces an angular acceleration.

(1) A net force F acting through the centre of gravity of a body of mass m causes an acceleration a in the direction of the force, given by Eqn. (2.1),

$$F = ma \dotfill (3.1)$$

(2) A net torque T acting on a body of moment of inertia I about the specified axis causes an angular acceleration a, given by Eqn. (2.2),

$$T = Ia \dotfill (3.2)$$

Collisions. Consider a simple collision of two spheres, masses m_1, m_2, travelling along their line of centres, and having velocities u_1, u_2 before impact, and v_1, v_2 after impact, respectively (Fig. 9).

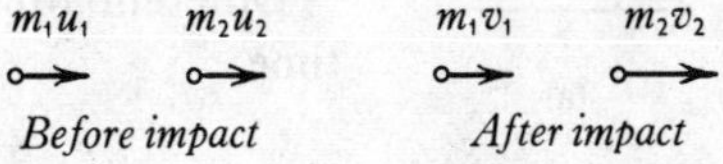

$m_1 u_1 \quad m_2 u_2 \qquad m_1 v_1 \quad m_2 v_2$

Before impact   *After impact*

Fig. 9.

Conservation of momentum applies to *all* collisions,

∴   Momentum before = Momentum after,

or $$m_1 u_1 + m_2 u_2 = m_1 v_1 + m_2 v_2 \dotfill (3.3)$$

The coefficient of restitution e is given by

$$e = \frac{\text{Relative velocity of separation}}{\text{Relative velocity of approach}} = \frac{v_2 - v_1}{u_1 - u_2} \dotfill (3.4)$$

Conservation of kinetic energy does not apply to the actual impact, since heat is produced, and the total kinetic energy afterwards is less than that before:

$$\tfrac{1}{2}m_1 u_1^2 + \tfrac{1}{2}m_2 u_2^2 = \tfrac{1}{2}m_1 v_1^2 + \tfrac{1}{2}m_2 v_2^2 + \text{Heat produced} \quad (3.5)$$

In a *perfectly elastic* collision, no kinetic energy would be lost. Rearranging Eqns. (3.3), (3.5) in this case,

$$m_1(u_1 - v_1) = m_2(v_2 - u_2),$$
$$m_1(u_1^2 - v_1^2) = m_2(v_2^2 - u_2^2),$$

11

and dividing one equation by the other, and cancelling,

$$u_1 + v_1 = v_2 + u_2,$$
or
$$u_1 - u_2 = v_2 - v_1.$$

Therefore $e = 1$ in this type of collision.

In practice, collisions are *imperfectly elastic*, the highest values of e being about 0.9 for hard steel, glass and ivory. An *inelastic* collision is one in which there is no rebound at all, $e = 0$, and maximum energy is converted to heat.

Radian measure. An angle θ in radian is defined (Fig. 10*a*) as $\dfrac{\text{Arc}}{\text{Radius}} = \dfrac{s}{r}$. A radian is thus the ratio of two lengths, and is dimensionless. Note also that 2π radian $= 360°$.

(a)

(b)

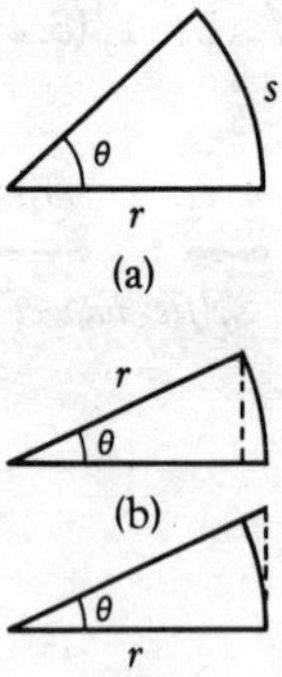

Fig. 10.

For small angles θ in radian (Fig. 10*b*),

$$\sin \theta \simeq \theta \qquad \text{and} \qquad \tan \theta \simeq \theta \quad \dots\dots\dots(3.6)$$

Differentiating the defining equation twice w.r.t. time,

$$\left.\begin{array}{l} s = r\theta, \\ v = r\omega, \\ a = ra. \end{array}\right\} \quad \dots\dots\dots\dots\dots(3.7)$$

We see that linear and angular displacements (s and θ), velocities (v and ω), and accelerations (a and a), are in each case related by a factor of r, when the angular measure is in radian.

Constant acceleration formulae. See Table (p. 16). Take care to apply these formulae only to cases of *constant* acceleration, e.g. a body falling freely under gravity, or acted upon by a constant force. In this latter case, if the constant force is F, the acceleration is given by $a = F/m$ (Eqn. 2.1).

The formulae can be applied to angular motions also, where angular units in radian replace linear units. In this case, a constant torque T produces an angular acceleration given by $a = T/I$ (Eqn. 2.2).

Worked examples

These should first be attempted by the student without reference to the working given. Always draw a clear diagram and explain working fully. Log tables are required only in Ex. 8.

3 . MECHANICS: PROBLEMS

1. *Friction on inclined planes.* What force, parallel to the plane, is required to cause a wooden block of mass 20 kg to move at uniform speed up a plane inclined at 30° to the horizontal, if the coefficient of sliding friction is 0.6?

2. *Power = Force × Speed.* What power is developed by a car of mass 1000 kg travelling at a steady speed of 15 m s^{-1} up a gradient of 1 in 10, if the frictional force is 400 N?

3. *Force of water jet.* What pressure does a horizontal jet of water travelling at 2 m s^{-1} exert on a vertical wall, assuming there is no rebound?

4. *Optimum angle of bank.* What is the optimum angle of bank of a railway track of radius of curvature 400 m, for a train travelling at 20 m s^{-1}?

5. *Hydrometers and flotation.* What is the cross-sectional area of the stem of a hydrometer, mass 0.015 kg, if the distance between the 1000 and 800 kg m^{-3} marks is 0.03 m?

6. *Tethered balloon.* What is the tension in the cable of a balloon, of volume 100 m³, mass of fabric 50 kg, filled with hydrogen? (Density of hydrogen = 0.09 kg m^{-3}; density of air = 1.29 kg m^{-3}.)

7. *Faulty barometers and Boyle's law.* A faulty barometer contains a small quantity of air in the space above the mercury. It reads 750 mmHg when the true atmospheric pressure is 760 mmHg, the air space then being 30 mm long. What will the true pressure be when the barometer reads 740 mmHg?

8. *Number of strokes of vacuum pump.* How many strokes of a vacuum pump, of effective volume 100 cm³, are required to reduce the pressure in a bell-jar of volume 900 cm³ to one tenth its initial value, assuming the air remains at constant temperature?

WORKING
  1. Resolving along plane (Fig. 11), $P = mg \sin \theta + \mu N$. Resolving perpendicular to plane, $N = mg \cos \theta$.

$$\therefore \qquad P = mg(\sin \theta + \mu \cos \theta)$$
$$= 20 \times 10 \, (\sin 30° + 0.6 \cos 30°)$$
$$= 204 \, \text{N}.$$

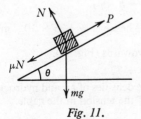

Fig. 11.

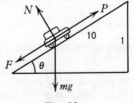

Fig. 12.

13

2. Resolving along plane (Fig. 12),

$$P = mg \sin \theta + F$$
$$= (1000 \times 10 \div 10) + 400$$
$$= 1400 \text{ N}.$$

∴ Power = Work done per second
= Force × Distance moved per second
= 1400 × 15
= 21 kW.

3. Pressure = Change of momentum per second on unit area of wall
= (Mass striking unit area per second) × (Its change of velocity)
= (Density × Volume arriving per second) × (Change of velocity).

If v is the velocity of impact, the volume arriving on unit area per second is numerically equal to v (Fig. 13).

∴ Pressure = $(1000 \times 2) \times (2) = 4000 \text{ N m}^{-2}$.

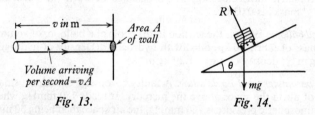

Fig. 13.   Fig. 14.

4. Optimum angle of bank means no tendency to slip outwards or inwards. Therefore the reaction R must be normal (Fig. 14).

Resolving vertically, $R \cos \theta = mg$.

The horizontal resolved force, $R \sin \theta$, is the resultant unbalanced force necessary to give the train an acceleration v^2/r towards the centre of curvature of the track. By Newton's second law, $R \sin \theta = mv^2/r$.

Dividing one equation by the other,

$$\tan \theta = \frac{v^2}{rg},$$

which is the required condition of bank.

Substituting, we obtain $\theta = 6°$ approximately.

5. Mass m of hydrometer = Mass of liquid displaced = (Volume of hydrometer immersed) × (Density of liquid).

Using the symbols in Fig. 15,

$$m = (V + h_1 A)\rho_1 = (V + h_2 A)\rho_2.$$

Eliminating V, $\dfrac{m}{\rho_2} - \dfrac{m}{\rho_1} = (h_2 - h_1)A$. Whence $A = 1.25 \times 10^{-4} \text{ m}^2$.

6. Equating upthrust to forces acting downwards (Fig. 16),

$$\rho_1 Vg = \rho_2 Vg + mg + T,$$

V being the volume of the balloon, ρ_1, ρ_2 the densities of air and hydrogen respectively, m the mass of the fabric, and T the tension in the cable. Whence $T = 700$ N.

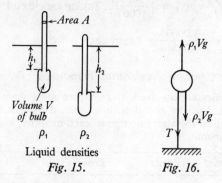

Liquid densities

Fig. 15.

Fig. 16.

7. In each case, True atmospheric pressure = Pressure of mercury column + Pressure in air space.

From Fig. 17, the length of the air column changes from 30 mm to 40 mm.

By Boyle's law, $P_2 = P_1 \dfrac{V_1}{V_2} = 10 \times \dfrac{30}{40} = 7.5$ mmHg.

Therefore, New atmospheric pressure = 740 + 7.5 = 747.5 mmHg.
Distinguish carefully between a length in mm and a pressure in mmHg.

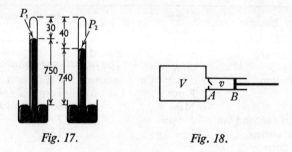

Fig. 17.   Fig. 18.

8. *Proof of formula.* At the first evacuating stroke (Fig. 18), as the piston moves from A to B the volume of air in the chamber increases from V to $(V + v)$.

By Boyle's law, $P_0 V = P_1(V + v)$, or $P_1 = P_0 \left(\dfrac{V}{V + v} \right)$.

At the second evacuating stroke the pressure P_1 is further reduced to P_2, given by $P_2 = P_1 \dfrac{V}{V + v} = P_0 \left(\dfrac{V}{V + v} \right)^2$.

Therefore after the nth stroke, $P_n = P_0 \left(\dfrac{V}{V + v} \right)^n$.

15

In this case, $\dfrac{P_n}{P_0} = 0.1 = \left(\dfrac{900}{1000}\right)^n$. Taking logs, $\log 0.1 = n \log 0.9$, and

$$n = \dfrac{\bar{1}.00000}{\bar{1}.95424} = \dfrac{-1.00000}{-0.04576} = 22 \text{ strokes.}$$

Table I—The Constant Acceleration Formulae

Five quantities are involved. If any three are known, the other two can be found. Use the equation that omits the quantity not required. Pay due regard to the algebraic sign of each quantity, and to consistency of units throughout.

Quantity omitted	Formula	Proof
Displacement, s	$v = u + at$	follows from definition of a
Acceleration, a	$s = \dfrac{u + v}{2}t$	Distance = Average velocity × Time
Final velocity, v	$s = ut + \frac{1}{2}at^2$	by substituting for v
Initial velocity, u	$s = vt - \frac{1}{2}at^2$	,, ,, ,, u
Time, t	$v^2 = u^2 + 2as$	,, ,, ,, t

It is essential that these formulae be known by heart.

Table II—Comparison of Linear and Angular Motions

	Linear motion		Angular motion
Displacement	s	$s = r\theta$	θ
Velocity	ds/dt or v	$v = r\omega$	$d\theta/dt$ or ω
Acceleration	d^2s/dt^2 or a	$a = ra$	$d^2\theta/dt^2$ or a
	Force F		Torque $T\,(= Fr)$
	Mass m		Moment of Inertia I
Newton's second law	$F = ma$		$T = Ia$
Kinetic energy	$\frac{1}{2}mv^2$		$\frac{1}{2}I\omega^2$
Work done	Fs		$T\theta$

4. Simple Harmonic Motion

Definitions

A PERIODIC motion is one which repeats itself exactly after constant time intervals.

The PERIOD (T) is the time for one complete oscillation.

The AMPLITUDE (a) is the maximum value of the displacement x from the centre of the oscillation.

SIMPLE HARMONIC MOTION is that periodic motion in which the acceleration of the particle is proportional to its displacement from, and always directed towards, a fixed point.

S.h.m. mathematically. Expressing the above definition of s.h.m. in mathematical form, $-\ddot{x} \propto x$. The minus sign indicates an acceleration towards, and not away from, the origin O. Thus, $-\ddot{x} = \omega^2 x$, where ω^2 is the constant of proportionality. The constant is made ω^2, and not ω, in order to ensure that it is a positive quantity, and also because ω so defined has the significance of *angular velocity* or *angular frequency*, as explained below.

Thus s.h.m. is represented by an equation of the form

$$\ddot{x} + \omega^2 x = 0 \quad\text{.............................(4.1)}$$

One solution of this equation is

$$x = a \cos \omega t \quad\text{.............................(4.2)}$$

where a is a constant representing the *amplitude*.

Differentiating twice w.r.t. time t,

$$\dot{x} = -\omega a \sin \omega t \quad\text{........................(4.3)}$$

and

$$\ddot{x} = -\omega^2 a \cos \omega t \quad\text{........................(4.4)}$$

showing that Eqn. (4.2) is a correct solution.

As the cosine repeats itself after an angle 2π, and in this case after an angle ωT, where T is the *period*, then $2\pi = \omega T$. So we have the important relation

$$T = \frac{2\pi}{\omega} \quad\text{.............................(4.5)}$$

S.h.m. as projection on a diameter of uniform circular motion. In Fig. 19, P' moves uniformly round the circle, radius a, with angular velocity ω in rad s^{-1}. Let the time t be zero when P' passes A'. Then the angle $A'OP'$ passed through by P' at time t is equal to ωt.

P is the projection of P' upon the fixed diameter AA'. Its displacement x from the centre O at time t is given by

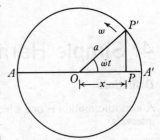

Fig. 19. *Uniform circular motion producing s.h.m.*

$$x = a \cos \omega t.$$

Therefore P executes s.h.m. along AA', with an amplitude a and a period of $T = \dfrac{2\pi}{\omega}$.

One useful feature of this approach is to show the significance of the constant ω and the angle ωt. The constant ω is the angular velocity of the particle in circular motion which generates s.h.m. along a diameter.

From Fig. 19, $\cos \omega t = \dfrac{x}{a}$, and $\sin \omega t = \dfrac{\surd(a^2 - x^2)}{a}$. It is sometimes useful to express the velocity $\dot{x}$ (Eqn. 4.3) in terms of the variable x instead of the variable t:

$$\dot{x} = -\omega \surd(a^2 - x^2) \quad\dots\dots\dots\dots\dots(4.6)$$

Material body executing s.h.m. If a body, mass m, is acted upon by a *restoring force F* which is always proportional to its displacement from, and directed towards, a fixed point, the body executes s.h.m.

Thus restoring force $F \propto$ displacement x,

or $$F = kx \dots\dots\dots\dots\dots\dots\dots\dots(4.7)$$

where $k = $ a constant $= $ *restoring force per unit displacement*.

Applying Newton's second law to the body, $F = -m\ddot{x}$, or $kx = -m\ddot{x}$. The minus sign indicates an acceleration towards, and not away from, the origin O.

$$\therefore \qquad \ddot{x} + \frac{k}{m}x = 0,$$

which is an equation of the form $\ddot{x} + \omega^2 x = 0$,

where $$\omega^2 = \frac{k}{m} \quad\dots\dots\dots\dots\dots\dots(4.8)$$

and is therefore s.h.m., having a period $T = 2\pi/\omega$,

18

or
$$T = 2\pi\sqrt{\frac{m}{k}} \dots\dots\dots\dots\dots\dots\dots (4.9)$$

Energy of body executing s.h.m.

Kinetic energy. The K.E. of the body at any instant $= \frac{1}{2}mv^2 = \frac{1}{2}m\dot{x}^2 = \frac{1}{2}m\omega^2a^2\sin^2\omega t$ (Eqn. 4.3). This reaches a maximum value of $\frac{1}{2}m\omega^2a^2$ as the body passes through the equilibrium position.

Potential energy. This is the work done against the restoring force F as the body moves from the equilibrium position. As $F(=kx)$ is not constant we must integrate. Total P.E. at displacement x

$$= \int_0^x F dx = \int_0^x kx dx = \frac{1}{2}kx^2 = \frac{1}{2}ka^2\cos^2\omega t = \frac{1}{2}m\omega^2a^2\cos^2\omega t$$

(Eqn. 4.8). This reaches a maximum value of $\frac{1}{2}m\omega^2a^2$ at the extremes of motion, and is zero at the centre.

Total energy. The sum of the K.E. and P.E. at any instant

$$= \frac{1}{2}m\omega^2a^2(\sin^2\omega t + \cos^2\omega t) = \frac{1}{2}m\omega^2a^2 = \text{a constant.}$$

Note also that the energy is proportional to the *square* of the amplitude; this is true of any oscillation or wave motion.

Summary of characteristics of s.h.m. We see that the definition of s.h.m. leads to a periodic motion of sinusoidal form, the oscillation being symmetrical about a fixed point, or equilibrium position, and of unvarying amplitude and periodic time. The two constants a and ω in the equation $x = a\cos\omega t$ determine the amplitude and period, respectively.

The motion can be produced on a body by a restoring force proportional to the displacement from, and directed towards, a fixed point.

The velocity of the particle reaches a maximum when the displacement and acceleration are zero, and vice versa (Fig. 20). The maximum values of these quantities are deduced from Eqns. (4.2–4).

The energy of the motion oscillates between the extremes of entirely potential and entirely kinetic energy, the total energy at any instant being constant. The energy is proportional to the square of the amplitude.

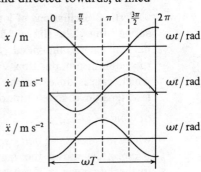

Fig. 20. Phase relationships in s.h.m.

Examples of s.h.m.

To show that a given motion is s.h.m. a procedure may be adopted as follows:—

Apply the appropriate physical principles—e.g. components of forces, Hooke's law, Archimedes' principle, pressure in a liquid—to find the restoring force for a given displacement x from the equilibrium position.

Apply Newton's second law to obtain the equation of motion.

If this equation falls into the form $\ddot{x} + \omega^2 x = 0$ the motion must be s.h.m. Furthermore, as a value has now been found for the constant ω^2 the period T can immediately be written down by substitution in the relation $T = 2\pi/\omega$.

The following well-known examples illustrate this procedure.

Small oscillations of simple pendulum. The instantaneous restoring force F is the component of mg perpendicular to the string (Fig. 21), i.e. $mg \sin \theta$.

Applying Newton's second law, $F = -m\ddot{x}$, or $mg \sin \theta = -m\ddot{x}$, where x is the displacement along the arc OP. For small angles θ, $\sin \theta \simeq \theta = x/l$. Substituting, we obtain

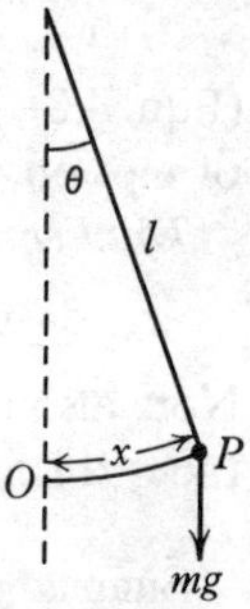

Fig. 21.

$$\ddot{x} + \frac{g}{l}x = 0.$$

This is of the form $\ddot{x} + \omega^2 x = 0$ and is therefore s.h.m., where $\omega^2 = g/l$, the period being

$$T = \frac{2\pi}{\omega} = 2\pi\sqrt{\frac{l}{g}} \quad\quad\quad\quad\quad (4.10)$$

Vertical oscillations of loaded, weightless spring. Fig. 22a shows the spring at its natural length, unloaded. Fig. 22b shows the equilibrium position with a load mg which causes an extension l. By Hooke's law, tension $\propto$ extension, or $mg \propto l$,

or $\quad mg = kl$ ....(4.11)

where $k = load\ per\ unit\ extension$.

Fig. 22c shows the mass displaced a further distance x from the equilibrium position. By Hooke's law again, tension $T' = k(l + x) = mg + kx$. Therefore re-

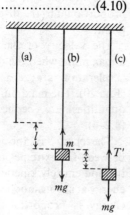

Fig. 22.

storing force upwards $= T' - mg = mg + kx - mg = kx$.

By Newton's second law, $F = -m\ddot{x}$, or $kx = -m\ddot{x}$,

or
$$\ddot{x} + \frac{k}{mx} = 0,$$

which is of the form $\ddot{x} + \omega^2 x = 0$ and is therefore s.h.m., with $\omega^2 = k/m = g/l$ (Eqn. 4.11). Thus period

$$T = \frac{2\pi}{\omega} = 2\pi\sqrt{\frac{m}{k}} = 2\pi\sqrt{\frac{l}{g}}\dots\dots\dots\dots(4.12)$$

Vertical oscillations of test-tube floating in non-viscous liquid. When the loaded test-tube is displaced a depth x beyond its equilibrium position (Fig. 23) it displaces an *additional* weight of liquid, which causes an upthrust, or restoring force, tending to return it to the equilibrium position. By Archimedes' principle the restoring force $F = mg = \rho Axg$, where m ($= \rho Ax$) is the additional mass of liquid displaced.

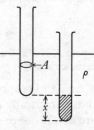

Fig. 23.

By Newton's second law, $F = -M\ddot{x}$, or $\rho Axg = -M\ddot{x}$, where M is the mass of the test-tube.

$\therefore$
$$\ddot{x} + \frac{\rho Ag}{M}x = 0$$

which is of the form $\ddot{x} + \omega^2 x = 0$ and is therefore s.h.m., with $\omega^2 = \rho Ag/M$, and period

$$T = \frac{2\pi}{\omega} = 2\pi\sqrt{\frac{M}{\rho Ag}}\dots\dots\dots\dots\dots(4.13)$$

Liquid oscillating in U-tube. Excess pressure on L.H.S. (Fig. 24) due to displacement x of *each* level $= h\rho g = 2x\rho g$, where $h = 2x$. The restoring force $= 2x\rho gA$, where A is the cross-sectional area of the U-tube.

By Newton's second law, $F = -m\ddot{x}$, or $2\rho gAx = -lA\rho\ddot{x}$, where m ($= lA\rho$) is the total mass of liquid in the U-tube, and l is the length of the liquid column.

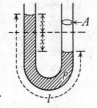

Fig. 24.

$\therefore$
$$\ddot{x} + \frac{2g}{l}x = 0.$$

This is of the form $\ddot{x} + \omega^2 x = 0$ and is therefore s.h.m., with $\omega^2 = 2g/l$, and period

$$T = \frac{2\pi}{\omega} = 2\pi\sqrt{\frac{l}{2g}}\dots\dots\dots\dots\dots(4.14)$$

21

5. Gravitation

Attraction between point masses

UNIVERSAL LAW OF GRAVITATION. Every particle of matter attracts every other particle with a force which varies directly as the product of their masses and inversely as the square of the distance between them.

The GRAVITATIONAL CONSTANT (G) is defined by the equation

$$F = G \frac{m_1 m_2}{r^2} \quad \ldots\ldots\ldots\ldots\ldots\ldots\ldots\ldots\ldots\ldots(5.1)$$

where F is the force of attraction between two point masses m_1, m_2 a distance apart r.

Eqn. (5.1) expresses the law of gravitation and defines G, which is thus numerically the force between two particles of unit mass which are unit distance apart.

$$G = 6.67 \times 10^{-11} \text{ N m}^2 \text{ kg}^{-2}$$

Homogeneous spherical masses. Eqn. (5.1) applies to point masses, but can also be applied to homogeneous spherical masses which in certain circumstances can be treated as point masses. It can be shown that:

(1) For points *outside* a spherical mass, e.g. point P (Fig. 25), the gravitational effect is as if the whole mass were concentrated at the centre O.

(2) For points *inside* the spherical mass, e.g. point Q, the gravitational effect is the same as if the outer, unshaded portion were removed, and the shaded matter were concentrated at O. In other words, the net effect of the unshaded portion on a particle at Q is nil.

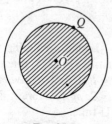

Fig. 25.

Variation of g with position above and below Earth's surface. This illustrates the facts noted in the previous two paragraphs.

(1) Value of g at a point P *above* the Earth. For a point outside the Earth the whole of the mass M of the Earth can be considered as acting at its centre. The gravitational force F on a body, mass m, a distance r from the centre of the Earth, is therefore given by

$$F = G \frac{Mm}{r^2} \quad \ldots\ldots\ldots\ldots\ldots\ldots\ldots\ldots(5.2)$$

But this gravitational force is by definition the weight W of the body,

$$\therefore \qquad F = W = mg \qquad\qquad\qquad\dots\dots\dots(5.3)$$

Therefore $mg = G\dfrac{Mm}{r^2}$, or $g \propto \dfrac{1}{r^2}$ at points above the Earth.

(2) Value of g at a point Q *inside* the Earth. In this case the effective mass of the Earth attracting the body is limited to that inside its own radius. Thus, if the mean density of the Earth is ρ and the body is a distance r from the centre of the Earth, the effective mass of the Earth is $\frac{4}{3}\pi r^3 \rho$ acting at its centre.

Therefore $mg = G\dfrac{\frac{4}{3}\pi r^3 \rho m}{r^2}$, or $g \propto r$ at points inside the Earth.

The conclusion is that the value of g is a maximum at the surface of the Earth, and decreases to zero at the centre.

Small satellite in circular orbit round Earth. By virtue of its circular motion, the satellite is at all times accelerating towards the centre of the Earth with an acceleration v^2/r. This means that a force of

$$F = m\frac{v^2}{r}\dots\dots\dots\dots\dots\dots\dots\dots\dots\dots\dots(5.4)$$

is required towards the centre to maintain the acceleration, where m is the mass of the satellite, v its linear velocity, and r the radius of its orbit. This force is provided by its own weight, the Earth's gravitational attraction:

$$F = mg' = G\frac{Mm}{r^2}\dots\dots\dots\dots\dots\dots\dots\dots(5.5)$$

where M is the mass of the Earth, and g' is the acceleration due to gravity at the height at which the satellite is.

Therefore $m\dfrac{v^2}{r} = G\dfrac{Mm}{r^2}$. If the satellite is in orbit close to the Earth, then r is also the Earth's radius. Taking r as 6.4×10^6 m and M as 6×10^{24} kg, we obtain for the velocity,

$$v = \sqrt{\frac{GM}{r}} = \sqrt{\frac{6.7 \times 10^{-11} \times 6 \times 10^{24}}{6.4 \times 10^6}} = 8 \times 10^3 \text{ m s}^{-1},$$

and for the period,

$$T = \frac{2\pi r}{v} = \frac{4 \times 10^7}{8 \times 10^3} = 5 \times 10^3 \text{ s} = 83 \text{ minute}.$$

To deduce Kepler's third law of planetary motion. Considering a circular orbit of a planet round the Sun, the law states that (Period)^2 $\propto$(Radius of orbit)^3. Newton first carried out this calculation, in showing that Kepler's laws could be explained by a single law of gravitation.

Assuming a circular orbit, where M is the mass of the Sun and m the mass of the planet, we can apply Eqns. (5.2) and (5.4), together with

$$T = \frac{2\pi r}{v} \dots\dots\dots\dots\dots\dots\dots\dots\dots(5.6)$$

giving

$$T^2 = \frac{4\pi^2 r^2}{v^2} = \frac{4\pi^2 r^3}{GM}, \quad \text{or} \quad T^2 \propto r^3.$$

6. Surface Tension

Origin of surface tension forces. Because of cohesive forces between its molecules, a liquid tends to assume the most compact form possible, i.e. to reduce its surface area to a minimum. It is therefore convenient to regard the liquid surface as if in some respects it were like an elastic membrane—tending to contract, and exerting pulling forces, tangential to the surface, on all neighbouring particles, and on the container at the points of contact.

The *angle of contact*, in this latter case, is important, and is determined by the relative magnitudes of the cohesive (liquid to liquid) and adhesive (liquid to container) forces. Compare the meniscus of water-in-glass with that of mercury-in-glass.

Definitions

The SURFACE TENSION (γ) is the force per unit length acting on either side of a line drawn in the surface of the liquid, in a direction perpendicular to the line and tangential to the surface. *Unit:* N m^{-1}.

Alternatively, the SURFACE TENSION of a liquid is the energy required per unit increase in surface area, under isothermic conditions. *Unit:* J m^{-2}.

These two definitions are equivalent (see below) and the units are identical, since $\dfrac{J}{m^2} = \dfrac{N\,m}{m^2} = \dfrac{N}{m}$.

The ANGLE OF CONTACT (θ) is the angle, inside the liquid, at which the liquid surface meets the solid surface.

To show that the two definitions are equivalent. $ABCD$ is a rectangular surface of a liquid, where $BC = b$ (Fig. 26). Neighbouring surfaces exert surface tension forces on the perimeter of $ABCD$ in the directions shown. By the first definition of surface tension T, in $N\,m^{-1}$, the force on BC is bT.

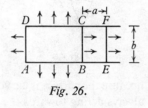

Fig. 26.

If we now extend BC a distance a to EF, the work done creating a new surface area ab is Force $\times$ Distance $= bTa$. This gives T, in $J\,m^{-2}$, which is in agreement with the second definition.

Excess pressure on concave side of spherical liquid surface. $ABCD$ is a small rectangular portion of a cylindrical liquid surface of radius r_1 (Fig. 27) and is bounded by sides of lengths a, b. Surface tension of liquid is γ.

The surface tension forces on AB, CD act in opposition and cancel out. So do the horizontal components of the forces on AD, BC. The vertical components of these latter forces, however, act in the same direction, and their resultant, of magnitude $2Tb \sin \theta$, acts towards the central axis O.

As θ is small, $\sin \theta \simeq \theta$. The force becomes $2\gamma b\theta$ or $\gamma ba/r_1$, where $2\theta = a/r_1$. This force acts over an area ab, therefore the surface tension *pressure* acting towards O is $\dfrac{\gamma}{r_1}$.

If the surface instead of being cylindrical is such that AD, BC also have a radius of curvature r_2, similar forces will be exerted on these two sides, resulting in an additional pressure of $\dfrac{\gamma}{r_2}$.

$\therefore$ $\qquad\qquad$ Total pressure $P = \gamma\left(\dfrac{1}{r_1} + \dfrac{1}{r_2}\right)$ ....(6.1)

For a single spherical surface, $r_1 = r_2$. Calling this r, the expression becomes

25

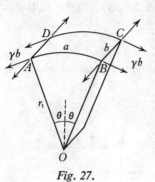

Fig. 27.

$$P = \frac{2\gamma}{r} \quad \dots\dots\dots\dots\dots\dots(6.2)$$

A spherical soap bubble has *two* surfaces. In this case therefore

$$P = \frac{4\gamma}{r} \quad \dots\dots\dots\dots\dots\dots(6.3)$$

which means that, for equilibrium, the air pressure inside the bubble must be in excess of that outside the bubble by this amount.

These surface tension pressures exist because the surface is curved and not flat. They are directed pressures, always acting towards the concave side of the curved liquid surface.

Capillary rise. The concave liquid meniscus is assumed to be spherical in shape (Fig. 28). R is the radius of the capillary tube at that point and r is the radius of the meniscus. The liquid, of surface tension γ and density ρ, makes an angle of contact θ with the glass. The liquid rises a height h up the tube. Barometric pressure is B.

The pressure at Y is made up of:—

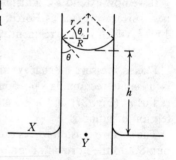

Fig. 28. Capillary rise.

Barometric pressure above meniscus $= B$
S.t. pressure (acting upwards) $= -2\gamma/r$
Pressure due to head of liquid $= h\rho g$.

But Pressure at $Y =$ Pressure at X at same level outside $= B$.

$$\therefore \quad B - \frac{2\gamma}{r} + h\rho g = B \quad \text{or} \quad \frac{2\gamma}{r} = h\rho g.$$

Also $R = r \cos \theta$.

$$\therefore \qquad \qquad \gamma = \frac{Rh\rho g}{2 \cos \theta} \quad \dots\dots\dots\dots\dots(6.4)$$

For water in clean glass, $\theta = 0$ and $\rho = 10^3 \text{ kg m}^{-3}$,

and
$$\gamma = \frac{Rh\rho g}{2} \quad \dots\dots\dots\dots\dots(6.5)$$

26

γ **by capillary rise.** Important points in this experiment include:
(1) preliminary cleaning of tube, glass scale and beaker;
(2) precautions when measuring h—e.g. ensuring that the meniscus rises freely in the tube when the beaker is raised; ensuring that there are no drops of liquid in the tube above the meniscus; repeating readings to check;
(3) the measurement of the radius of the tube—by travelling microscope or by weighing and measuring a thread of mercury; and
(4) taking the temperature of the liquid, since surface tension varies with temperature.

7. Elasticity

Types of elasticity. *Elasticity* is the ability of a body to regain its original shape when deforming forces on it are removed; the opposite of *elastic* is *plastic*. There are three types of deformation from which a body can recover—stretching (or compressing), twisting, and volume changes. Each type of deformation has its own elastic modulus; these are respectively Young's modulus, the rigidity modulus, and the bulk modulus. We are concerned here only with Young's modulus.

Definitions

STRESS = Deforming force per unit area.

STRAIN = Resulting fractional change in dimension.

A MODULUS OF ELASTICITY is the quotient $\dfrac{\text{Stress}}{\text{Strain}}$ for the given type of the deformation applied.

YOUNG'S MODULUS $= \dfrac{\text{Longitudinal stress}}{\text{Longitudinal strain}}$. *Unit:* N m^{-2}.

HOOKE'S LAW. Provided that the elastic limit is not exceeded, the extension of a wire is proportional to the applied tension.

Load-extension graphs for wires. For a wire subjected to a stretching force four points on the load-extension graph can be distinguished, in this order:—

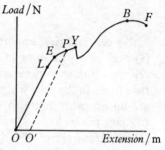

(1) The *limit of proportionality*, *L*, beyond which Hooke's law is not obeyed (Fig. 29).

(2) The *elastic limit*, *E*, the point beyond which the wire does not return to its original length when unloaded, but acquires a permanent extension. These first two points lie close together but do not quite coincide.

Fig. 29. Load-extension graph for steel.

On unloading at a point *P*, say, beyond the elastic limit, the wire recovers along the path *PO'*.

(3) The *yield point*, at *Y*, is the value of the load at which *plastic flow* begins, the wire continuing to stretch for some time after the load is applied, thinning uniformly along its length. Beyond this point the wire thus exhibits the property of *ductility*.

For steel, an actual drop in the load, as the material expands quickly, is recorded at *Y* (as in Fig. 29) when a special self-recording testing apparatus is used. This kink in the graph indicates a sudden change in internal arrangement of the molecules, resulting in a momentary weakness of the structure. Other metals do not exhibit such a well-defined yield point but otherwise the graphs are similar.

(4) Finally, a local thinning precedes fracture, at *F*. The *tensile strength* or *breaking stress*, exerted at *B*, is defined as the maximum stress the specimen can sustain without fracturing.

Young's modulus formula. Young's modulus *E* is measured within the region *OL*, in which the value of Stress/Strain remains constant (Fig. 29). A wire of natural length *L* and cross-sectional area *A* is loaded with a mass *m* resulting in an extension *l*. From these measured values *E* can be calculated:

$$E = \frac{\text{Stress}}{\text{Strain}} = \frac{F/A}{l/L} = \frac{FL}{Al} = \frac{mgL}{Al} \quad \ldots\ldots\ldots\ldots(7.1)$$

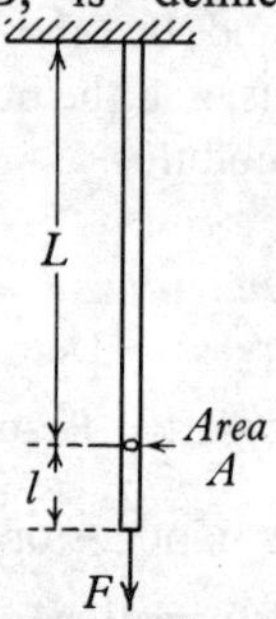

Fig. 30.

Potential energy stored in strained wire. This is the work done stretching the wire against the tension. The tension is not constant during stretching: for a given extension *x* the tension *F* in the wire is given by Eqn. (7.1),

$$F = \frac{EA}{L}x = kx \quad \ldots\ldots\ldots\ldots\ldots\ldots(7.2)$$

where *k* = a constant = *tension per unit extension*.

28

Total P.E. stored in stretching to an extension l is thus

$$\int_{x=0}^{l} F dx = \int_{x=0}^{l} kx\,dx = \left[\tfrac{1}{2}kx^2\right]_{0}^{l} = \tfrac{1}{2}kl^2 = \tfrac{1}{2}.kl.l$$

or $\qquad$ Total P.E. $= \tfrac{1}{2}$. Final tension . Extension $\quad$ ....(7.3)

As volume of wire is LA, for unit volume we have $\dfrac{\tfrac{1}{2}kl^2}{LA}$ or $\tfrac{1}{2} . \dfrac{kl}{A} . \dfrac{l}{L}$

or $\qquad$ P.E. stored per unit volume $= \tfrac{1}{2}$. Final stress . Strain ...(7.4)

If the work is done by hanging a weight on the wire, the gain in potential energy by the wire when the equilibrium position is reached is only half that lost by the weight, as Eqn. (7.3) shows. How is the balance of energy accounted for?

Tension in wire which is heated, clamped, and allowed to cool. A wire of length L at temperature θ_1 is heated to θ_2, clamped, and allowed to cool to its former temperature. If Young's modulus is E, linear expansivity is a, and cross-sectional area is A, what tension F results in the wire?

WORKING.—Increase in length l of wire when heated is given by the expansion formula (Eqn. 22.1),

$$l = La(\theta_2 - \theta_1).$$

When the wire has cooled, this increase is maintained by a tension F given by Eqn. (7.1),

$$F = \frac{EAl}{L}.$$

Substituting for l,

$$F = EAa(\theta_2 - \theta_1).$$

The tension is independent of the length.

8. Experiments

Ballistic balance. The scale pans are suspended by sets of cords so that they move over the arc of a circle of large radius and have no rotational movement as they swing. The masses added should not be free to slide about. Pointers are provided to indicate horizontal displacements (Fig. 31).

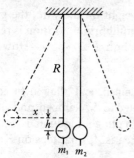

Fig. 31. Ballistic balance.

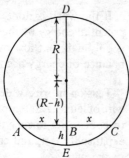

Fig. 32.

At the actual impact, momentum is conserved:

$$m_1 u_1 = m_1 v_1 + m_2 v_2 \quad\quad\quad\quad\quad\quad (8.1)$$

Instead of measuring the velocities u_1, v_1, v_2 of the masses just before or after impact, we measure the horizontal displacements x_1, y_1, y_2 to which they correspond. We can show that the velocity v of a mass m is proportional to its horizontal displacement x by applying the energy equation to its swing (Fig. 31):

$$\tfrac{1}{2}mv^2 = mgh,$$

where, by geometry,* $h = \dfrac{x^2}{2R}$. Therefore $v^2 = 2gh = \dfrac{gx^2}{R}$,

so $v \propto x$, or $\quad\quad\quad\quad\quad\quad v = kx \quad\quad\quad\quad\quad\quad\quad\quad (8.2)$

Substituting $u_1 = kx_1$, $v_1 = ky_1$, $v_2 = ky_2$ in Eqn. (8.1) we obtain,

$$m_1 x_1 = m_1 y_1 + m_2 y_2 \quad\quad\quad\quad\quad\quad (8.3)$$

Uses. (1) To verify the law of conservation of momentum. Eqn. (8.3) can be verified directly by measurement of all the quantities.

* By geometry (Fig. 32), $AB \cdot BC = DB \cdot BE$, or $x^2 = (2R-h)h$. Since $h \ll R$. we may write $x^2 = 2Rh$.

30

(2) To compare two masses. Rearranging Eqn. (8.3),

$$\frac{m_1}{m_2} = \frac{y_2}{x_1 - y_1} \qquad \dots\dots\dots\dots\dots(8.4)$$

(3) To find the coefficient of restitution between two bodies.

$$e = \frac{v_2 - v_1}{u_1} = \frac{y_2 - y_1}{x_1} \qquad \dots\dots\dots\dots(8.5)$$

Fletcher's frictionless trolley. The trolley (Fig. 33) is mounted on light wheels. A long spring, clamped at one end, carries an inked brush at the other. The brush touches lightly on a strip of paper fixed on the trolley. The spring is set in vibration, and, as the trolley moves, a wavy trace is drawn on the paper. Knowing the frequency of the vibrating spring, the acceleration of the trolley can be calculated from this trace.

Friction is 'eliminated' by raising the end of the trolley base until the trolley runs at a constant speed of its own accord, as indicated by an evenly spaced trace.

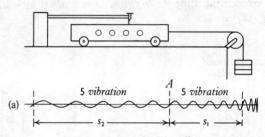

Fig. 33. Fletcher's trolley.

In the experiment the three quantities in Eqn. (2.1) can be measured. F is mg, where m is the mass of the scale pan and weights. M is the mass of the trolley itself plus scale pan and weights. The acceleration a is found by measuring s_1 and s_2 (Fig. 33a) for five complete vibrations each, and calculating a as follows:

$$s_1 = vt - \tfrac{1}{2}at^2,$$

where v is the 'final' velocity, at A on the trace, and

$$s_2 = ut + \tfrac{1}{2}at^2,$$

where u is the 'initial' velocity, at A on the trace.
But $u = v$. Subtracting the first equation from the second,

$$s_2 - s_1 = at^2 \qquad \dots\dots\dots\dots\dots(8.6)$$

31

The time t of five complete oscillation is known from a separate experiment in which the oscillation of the brush is counted and timed. Hence a can be calculated from Eqn. (8.6).

The validity of the formula $F = Ma$ can thus be tested by showing (1) that $a \propto F$, when M is kept constant; and (2) that $a \propto \dfrac{1}{M}$, when the mass of the trolley is altered, keeping F constant. In the first part of the experiment, such weights as will later be used as 'force weights' in the scale pan should meanwhile ride as passengers on the trolley, so that the total inertia mass M remains constant.

Determination of Young's modulus for a wire. Readings of load m against extension l are taken as the load is increased, in steps of say 1 kg, up to a point well below the elastic limit. The extension is read on a vernier scale, or by Searle's apparatus. Readings are then taken as the load is decreased. A mean load-extension graph is plotted. The diameter of the wire is measured in several places with a screw gauge and the average cross-sectional area A calculated. The length L of the wire from the support *to the vernier scale* is measured with a ruler (Fig. 34).

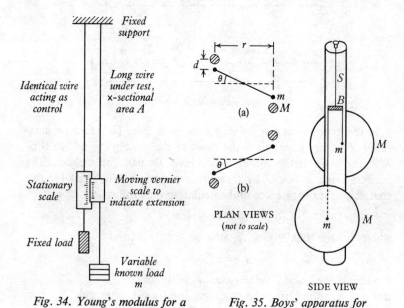

Fig. 34. Young's modulus for a wire.

Fig. 35. Boys' apparatus for determination of G.

Writing Eqn. (7.1) in the form of a straight line graph $y = kx$,

$$l = \frac{gL}{AE}m \quad \dots\dots\dots\dots\dots\dots(8.7)$$

The slope k of the straight part of the load-extension graph (with m as the x-axis) is therefore gL/AE. The numerical value found from the slope of the graph plotted is equated with this quantity and E, the only unknown, is calculated.

This method with two identical wires eliminates two sources of error—yielding of the support, and expansion due to temperature changes. These affect both wires equally.

It should be noted that in terms of percentage error in the final result an error of 1 cm in measuring the length L is less important than an error of 0.001 cm in measuring the radius.

Cavendish's torsion method for G. In Boys' modified apparatus (Fig. 35) a horizontal mirror B was suspended from a fine quartz torsion fibre S. From the ends of the mirror were suspended two gold spheres m, m at different levels, the whole being enclosed in an evacuated glass tube to prevent draughts. Two lead spheres M, M were placed as close as possible to the gold spheres, in such a position that the gravitational attraction between each lead sphere and its adjacent gold sphere caused a torque and a resulting angle of twist θ in the quartz suspension. This deflection θ was observed through a telescope trained on the usual lamp-and-scale device.

The lead spheres were then moved to a position giving a similar deflection in the opposite sense. The total angle measured was thus 2θ (Fig. 35a, b).

Equating the torsional and gravitational torques,

$$c\theta = G\frac{Mm}{d^2}r,$$

where c is the torsional torque per unit angle of twist in the quartz. The quantity c is found by a separate experiment in which the period T of free swing of the suspended system about the axis S is measured. This is given by

$$T = 2\pi\sqrt{\frac{I}{c}},$$

where I is the known moment of inertia of the system. Hence G is calculated

Extreme precautions to avoid vibration were taken. The different heights of the gold spheres avoided appreciable attractions between the lower lead sphere and upper gold sphere and vice versa.

The success of the experiment depends on the unusual properties of the quartz fibre, which has sufficient tensile strength to support the gold masses, but offers a small enough resistance to torsion for θ to be large enough to be measurable. Boys made the dimensions of the apparatus small (e.g. r about 2.5 cm) because, although a large r increases the gravitational torque, it also increases I, making T excessive. A period of swing of about 5 minute was used.

Part II
Light

9. Reflection and Refraction

Ray diagrams. A clear ray diagram often provides the key to a problem. The accepted conventions should always be followed when drawing ray diagrams—otherwise the diagram may not help, and it may mislead. The important conventions include: (1) arrows on all rays; (2) broken lines for rays produced back to a virtual image or forward to a virtual object; (3) object labelled O, image labelled I; (4) shaded backing to a line representing a mirror surface; (5) arrows in *both* directions on rays which are reflected normally from some surface to return along the same path; and (6) a right angle marked clearly at the surface from which such rays are reflected.

When producing rays backwards or forwards with broken lines, make sure that they really *are* produced, in the same straight line. When a ray bends at a change of medium, show the bend the correct way—towards the normal when entering the more dense medium, away from the normal when entering the less dense medium. Do not forget the fairly obvious fact that light rays travel *towards* the observer's eye. In lens and mirror problems, use a point object on the axis rather than an extended object, if possible. Be prepared, however, to draw constructional rays from extended objects if necessary, and also continuous pencils of light through optical instruments, from points off the axis.

Reflection

LAWS OF REFLECTION. (1) The incident ray, normal, and reflected ray all lie in the same plane.

(2) The angle of incidence i is equal to the angle of reflection r (Fig. 36).*

The image in a plane mirror is located as far behind the mirror as the object is in front, and lies on the same normal.

Two mirrors form a number of images of a single object, on the principles that an image in one mirror acts as an object for the other mirror. In the example shown (Fig. 37), with the two mirrors at right angles, the image I_1 in mirror AB acts as an object for mirror BC, forming an image I_3. Similarly the image I_2 acts as an object for AB, forming another image coincident with I_3.

* Note that all angles i and r are measured between ray and *normal*.

9 . REFLECTION AND REFRACTION

A reflected ray turns through twice the angle turned through by the mirror. In Fig. 38, an incident ray PQ falling normally on mirror AB is reflected back along the same path. When the mirror is rotated through an angle θ to position CD the reflected ray lies along QR, i.e. has rotated through 2θ.

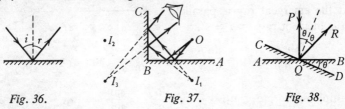

Fig. 36.   *Fig. 37.*   *Fig. 38.*

Refraction

LAWS OF REFRACTION. (1) The incident ray, normal, and refracted ray all lie in the same plane.

(2) *Snell's law.* For the two given media and a particular colour, the ratio $\sin i/\sin r$ is constant, where i and r are the angles of incidence and refraction (Fig. 39).*

The REFRACTIVE INDEX (n) is the constant given by

$$n = \frac{\sin i}{\sin r} \quad\text{.................................(9.1)}$$

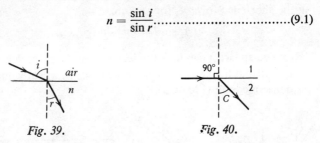

Fig. 39.   *Fig. 40.*

Total internal reflection

The CRITICAL ANGLE (c) is the greatest angle of incidence, for a ray passing from the more dense to the less dense medium, for which refraction can occur.

TOTAL INTERNAL REFLECTION occurs when a ray, passing from a more dense to a less dense medium, strikes the surface at an angle of incidence greater than the critical angle.

When a ray from medium 1 (Fig. 40) enters a more dense medium 2 at grazing incidence, $i = 90°$ and $r = c$. Substituting in Eqn. (9.1),

$$_1n_2 = \frac{1}{\sin c} \quad\text{..............................(9.2)}$$

* Note that all angles i and r are measured between ray and *normal*.

LIGHT

Refractive index between any two media. *AB, CD, EF* are three parallel plane boundaries between media as shown (Fig. 41). In such a case it is an experimental fact that the emergent ray is parallel to the incident ray, or $a = a'$.

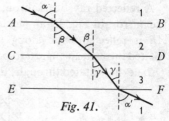

Fig. 41.

$$\therefore \quad {}_1n_2 = \frac{\sin a}{\sin \beta}, \quad {}_2n_3 = \frac{\sin \beta}{\sin \gamma}, \quad {}_3n_1 = \frac{\sin \gamma}{\sin a}, \quad \text{and} \quad {}_1n_3 = \frac{\sin a}{\sin \gamma}.$$

Multiplying, $\quad {}_1n_2 \cdot {}_2n_3 = \frac{\sin a}{\sin \beta} \cdot \frac{\sin \beta}{\sin \gamma} = \frac{\sin a}{\sin \gamma} = {}_1n_3.$

$$\therefore \qquad\qquad\qquad {}_1n_3 = {}_1n_2 \cdot {}_2n_3 \quad\dots\dots\dots\dots\dots\dots(9.3)$$

General refraction formula. A more general form of Eqn. (9.1) is sometimes more convenient. For a ray passing between media 1 and 2 (Fig. 42) ${}_1n_2 = \frac{\sin i_1}{\sin i_2}$. But ${}_1n_2 = {}_1n_a \cdot {}_an_2 = \frac{{}_an_2}{{}_an_1} = \frac{n_2}{n_1}$, writing the shortened form n_1 for ${}_an_1$, etc.

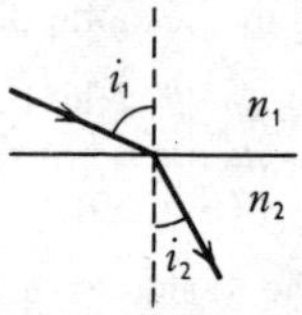

Fig. 42.

$$\therefore \qquad\qquad\qquad n_1 \sin i_1 = n_2 \sin i_2 \quad\dots\dots\dots\dots\dots\dots(9.4)$$

Real and apparent depth. An object at O in the more dense medium forms a virtual image at I, when viewed from vertically above in the less dense medium (Fig. 43a).

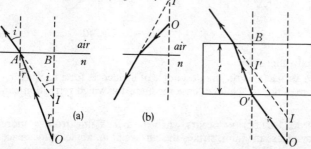

Fig. 43. Real and apparent depth.   *Fig. 44. Displacement towards observer.*

Now $\qquad n = \frac{\sin i}{\sin r}, \qquad \sin i = \frac{AB}{AI}, \qquad \sin r = \frac{AB}{AO}.$

Substituting, $\qquad\qquad n = \frac{AB/AI}{AB/AO} = \frac{AO}{AI}.$

9 . REFLECTION AND REFRACTION

When viewed normally, i and r are small, and $AO \simeq BO$, $AI \simeq BI$.

$$\therefore \qquad n = \frac{BO}{BI} = \frac{\text{Real depth}}{\text{Apparent depth}} \qquad \dots\dots\dots\dots(9.5)$$

This result can be adapted to the case of rays passing in the reverse direction (Fig. 43b), provided that the object is being viewed normally.

Displacement towards observer. This is an extension of Eqn. (9.5). When a glass block of thickness d and refractive index n is interposed between observer and object (Fig. 44) the apparent displacement towards the observer of the object, when viewed normally, is

$$IO = I'O' = BO' - BI' = d - \frac{d}{n} \text{ (since } n = \frac{n}{BI'}, \text{ Eqn. 9.5).}$$

$$\therefore \qquad \text{Displacement towards observer} = d - \frac{d}{n} \dots\dots\dots(9.6)$$

and the displacement is independent of the position of the glass block.

Minimum deviation through a prism. Since light rays are reversible there must be, in general, *two* angles of incidence, i_1 and i_2, that give the same angle of deviation D (Fig. 45a). These two values of i should coincide where D reached a maximum or minimum value.

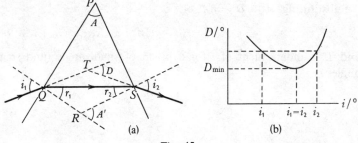

Fig. 45.

Experimentally, D does in fact reach a *minimum* value (Fig. 45b). The angle of minimum deviation $D_{\min}$ therefore occurs when $i_1 = i_2$ and $r_1 = r_2$; that is, when the ray passes symmetrically through the prism.

$PQRS$ is a cyclic quadrilateral, so Angle A (the refracting angle of the prism) = Angle A'.

39

Refractive index $n = \dfrac{\sin i_1}{\sin r_1} = \dfrac{\sin i_2}{\sin r_2}$. At minimum deviation these ratios are the same, and we can write (calling the angles i and r)

$$n = \frac{\sin i}{\sin r}.$$

In the quadrilateral $QRST$, Interior angles $i_1 + i_2 =$ Exterior angles $A + D$. At minimum deviation, this becomes $2i = A + D_{min}$. In the triangle QRS, $r_1 + r_2 = A$. At minimum deviation, $2r = A$. Substituting these values of i and r,

$$n = \frac{\sin \frac{1}{2}(A + D_{min})}{\sin \frac{1}{2}A} \quad \ldots\ldots\ldots\ldots\ldots\ldots(9.7)$$

Prism of small angle. Still using Fig. 45a, but where A is now small. For a prism of small angle we consider a more general case, not the particular case of minimum deviation.

As before, $A = A'$.

Also, $n = \dfrac{\sin i_1}{\sin r_1} = \dfrac{\sin i_2}{\sin r_2}$. But we can write $n \simeq \dfrac{i_1}{r_1} \simeq \dfrac{i_2}{r_2}$, if incidence

is near to normal, so that all angles are small.

$$\therefore \qquad\qquad i_1 + i_2 = n(r_1 + r_2).$$

As before, $i_1 + i_2 = A + D$; and $r_1 + r_2 = A$.
Substituting, $A + D = nA$, or

$$D = (n - 1)A \quad \ldots\ldots\ldots\ldots\ldots\ldots(9.8)$$

and D is independent of i for a prism of small angle (and small angles i).

10. Lenses and Mirrors

Sign conventions. §§ 10–13 involve the use of a sign convention. Two alternative conventions are given—*Real-is-Positive* (**RP**) and *New Cartesian* (**NC**). It is only necessary to know one of these, but the convention chosen must be understood thoroughly and applied consistently.

The purpose of a sign convention is to allow a given formula to be applied always in the same form, no matter where the object or image may be located, or what type of lens or mirror is used. In any calculation, therefore, the formula should first be put down, as it stands, in its algebraic form. It is only subsequently—when substituting *numerical values*—that the $+$ and $-$ signs are introduced.

Real-is-Positive convention. It is best to consider separately the conventions for u and v, f and r:

Object and image distances, u and v, are positive for real objects and images, negative for virtual.

Focal lengths f are positive for converging systems (convex lenses, concave mirrors), negative for diverging systems (concave lenses, convex mirrors).

Radius of curvature r. In the case of spherical mirrors, r follows the same convention as f, since $r = 2f$ (Eqn. 10.5). In the case of spherical refracting surfaces, including lens surfaces, r is *positive when the surface is convex to the less dense medium*, and otherwise negative.

In addition, for consistency in all possible cases, Eqn. (10.1) must be written as shown, where $n_2 \sim n_1$ means 'the positive value of $n_2 - n_1$'.

New Cartesian convention. This is the 'graph paper' convention:

All distances u, v, f, r measured to the right of the lens or mirror are positive, all distances to the left are negative.

The direction of the incident light is from left to right.

In the present notes no convention is used for *magnification*—all magnifications are considered to be positive.

Definitions

The AXIS of a lens is the line joining the centres of curvature of the two surfaces.

The OPTICAL CENTRE of a lens is the point through which any ray passing is undeviated.

The FOCAL POINT OF A CONVEX LENS (or 'Focus', 'Principal focus') is the point towards which rays *parallel to the axis* converge after passing through the lens.

The FOCAL POINT OF A CONCAVE LENS is the point from which rays parallel to the axis appear to diverge after passing through the lens.

The FOCAL LENGTH of a lens is the distance from the optical centre to the focal point.

41

A REAL IMAGE is an image through which the rays actually pass. It can be focused on a screen.

A VIRTUAL IMAGE is an image from which the rays appear to diverge. Since the rays do not actually pass through a virtual image it cannot be focused on a screen.

A REAL OBJECT is an object from which the rays diverge.

A VIRTUAL OBJECT is a point towards which the rays appear to be converging when intercepted by the lens or mirror.

Refraction at a single spherical surface. A real image I of the object O is formed by the surface of radius r. The light is entering the medium of higher refractive index n_2 (Fig. 46).

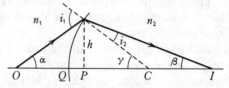

As all angles are small, sines and tangents can be approximated to angles in

Fig. 46. Spherical refracting surface.

radian, and points P, Q can be considered coincident.

Thus $\qquad n_1 \sin i_1 = n_2 \sin i_2$, $\quad$ or $\quad$ $n_1 i_1 \simeq n_2 i_2$.

Also $\qquad i_1 = a + \gamma$ $\quad$ and $\quad$ $i_2 = \gamma - \beta$.

But $\quad a \simeq \tan a = \dfrac{h}{PO}$, $\quad \beta \simeq \tan \beta = \dfrac{h}{PI}$, $\quad \gamma \simeq \tan \gamma = \dfrac{h}{PC}$.

Substituting, $\qquad n_1(a + \gamma) = n_2(\gamma - \beta)$,

or $\qquad n_1\left(\dfrac{h}{PO} + \dfrac{h}{PC}\right) = n_2\left(\dfrac{h}{PC} - \dfrac{h}{PI}\right)$.

NC	RP
$u = -PO, v = PI, r = PC.$	$u = PO, v = PI, r = PC.$
$\dfrac{n_2}{v} - \dfrac{n_1}{u} = \dfrac{n_2 - n_1}{r}$	$\therefore \dfrac{n_1}{u} + \dfrac{n_2}{v} = \dfrac{n_2 \sim n_1}{r}$ ...(10.1)
	See note (p. 41) for meaning of $n_2 \sim n_1$.

Thin lens formula. Two alternative derivations are given. The deviation method does not involve the use of the formula for refraction at a single spherical surface.

42

A. *Considering two spherical refracting surfaces.* A thin bi-convex lens (n_2) in a medium (n_1), where $n_2 > n_1$, forms a real image I_2 of a real object O (Fig. 47). As

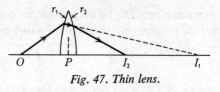

Fig. 47. Thin lens.

the lens is thin all measurements can be taken from its optical centre.

The first surface forms a real image I_1. I_1 acts as a *virtual object* for the second surface which forms a real image I_2. Taking each surface in turn,

NC	RP

NC:
$$\frac{n_2}{PI_1} - \frac{n_1}{u} = \frac{n_2 - n_1}{r_1},$$

$$\frac{n_1}{v} - \frac{n_2}{PI_1} = \frac{n_1 - n_2}{r_2}.$$

RP:
$$\frac{n_1}{u} + \frac{n_2}{PI_1} = \frac{n_2 - n_1}{r_1},$$

$$\frac{n_2}{-PI_1} + \frac{n_1}{v} = \frac{n_2 - n_1}{r_2}.$$

Adding (NC),
$$\frac{n_1}{v} - \frac{n_1}{u} = (n_2 - n_1)\left(\frac{1}{r_1} - \frac{1}{r_2}\right).$$

Adding (RP),
$$\frac{n_1}{u} + \frac{n_1}{v} = (n_2 - n_1)\left(\frac{1}{r_1} + \frac{1}{r_2}\right).$$

For a thin lens in air (NC),
$$\frac{1}{v} - \frac{1}{u} = (n - 1)\left(\frac{1}{r_1} - \frac{1}{r_2}\right).$$

For a thin lens in air (RP),
$$\frac{1}{u} + \frac{1}{v} = (n - 1)\left(\frac{1}{r_1} + \frac{1}{r_2}\right).$$

When $u = \infty$, then $v = f$ (by definition of f). Substituting these values,

NC:
$$\frac{1}{f} = (n - 1)\left(\frac{1}{r_1} - \frac{1}{r_2}\right).$$

RP:
$$\frac{1}{f} = (n - 1)\left(\frac{1}{r_1} + \frac{1}{r_2}\right).$$

Thus
$$\frac{1}{f} = \frac{1}{u} + \frac{1}{v} = (n - 1)\left(\frac{1}{r_1} + \frac{1}{r_2}\right) \quad \text{RP}$$

$$\frac{1}{f} = \frac{1}{v} - \frac{1}{u} = (n - 1)\left(\frac{1}{r_1} - \frac{1}{r_2}\right) \quad \text{NC}$$

$$\left.\begin{array}{c} \\ \\ \end{array}\right\} \quad \ldots\ldots\ldots\ldots(10.2)$$

B. *Deviation method.* We consider a thin lens in air. The two lens surfaces at a distance h from the optical centre are considered to act as a prism of small angle A.

In Fig. 48a, C_1, C_2 are the centres of curvature of the two lens surfaces, and A' is the angle between the normals to the prism faces at the points of contact. Therefore $A = A'$. As the lens is thin all measurements can be taken from its optical centre. As all angles are small, tangents can be approximated to angles in radian.

$$\therefore \qquad A = A' = \phi + \theta = \frac{h}{PC_1} + \frac{h}{PC_2}.$$

In Fig. 48b, a ray parallel to the axis is deviated through an angle D to the focal point, where D is given by Eqn. (9.8),

$$D = (n - 1)\, A.$$

Also $$D = \frac{h}{PF}.$$

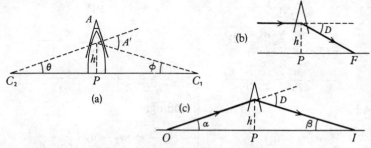

Fig. 48. *Thin lens. Deviation method.*

In Fig. 48c, a real image I is formed of a real object O, the ray shown being deviated through the same angle D as before.

$$\therefore \qquad D = a + \beta = \frac{h}{PO} + \frac{h}{PI}.$$

Combining all these equations,

$$\frac{h}{PF} = \frac{h}{PO} + \frac{h}{PI} = (n - 1)\left(\frac{h}{PC_1} + \frac{h}{PC_2}\right).$$

NC	RP
$f = PF, u = -PO, v = PI,$	$f = PF, u = PO, v = PI,$
$r_1 = PC_1, r_2 = -PC_2,$	$r_1 = PC_1, r_2 = PC_2,$

and we obtain the respective Eqns. (10.2) above.

Linear magnification formula. By definition, Linear magnification $m = \dfrac{\text{Size of image}}{\text{Size of object}} = \dfrac{I}{O}$ (Fig. 49). But, by similar triangles, $\dfrac{I}{O} = \left|\dfrac{v}{u}\right|$.

$$\therefore \qquad m = \frac{I}{O} = \left|\frac{v}{u}\right| \qquad \dots\dots\dots\dots(10.3)$$

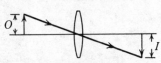

Fig. 49. Magnification.

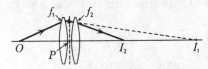

Fig. 50. Lenses in contact.

This important relation applies to all lenses (in air) and spherical mirrors.

Minimum separation of object and real image. For a convex lens $\frac{1}{f} = \frac{1}{U} + \frac{1}{D-U}$ (Eqn. 10.2), where U is the positive object distance and D is the total distance between object and *real* image.

Simplifying, $U^2 - DU + Df = 0.$

This is a quadratic equation having two solutions for U for a given value of D,

i.e. $$U = \frac{D \pm \sqrt{D^2 - 4Df}}{2}.$$

For real roots $D^2 \geqslant 4Df$, or $D \geqslant 4f$ (since $D \neq 0$). The minimum value of D is therefore $4f$, and from the solution it is seen that this occurs when $U = 2f$, that is, when object and image are symmetrically placed with respect to the lens (see Fig. 68).

Two thin lenses in contact. Two thin lenses of focal lengths f_1 and f_2, in contact, form a real image I_2 of a real object O (Fig. 50). As the lenses are thin all measurements can be taken from the centre of the system.

The first lens forms a real image I_1. I_1 acts as a *virtual object* for the second lens, which forms a real image I_2. Taking each lens in turn,

NC	RP
$$\frac{1}{PI_1} - \frac{1}{u} = \frac{1}{f_1},$$	$$\frac{1}{u} + \frac{1}{PI_1} = \frac{1}{f_1},$$
$$\frac{1}{v} - \frac{1}{PI_1} = \frac{1}{f_2}.$$	$$\frac{1}{-PI_1} + \frac{1}{v} = \frac{1}{f_2}.$$

Adding, Adding,

$$\frac{1}{v} - \frac{1}{u} = \frac{1}{f_1} + \frac{1}{f_2}.$$ $$\frac{1}{u} + \frac{1}{v} = \frac{1}{f_1} + \frac{1}{f_2}.$$

45

Considering the system as a single lens of focal length F,

$$\frac{1}{v} - \frac{1}{u} = \frac{1}{F}. \qquad\qquad \frac{1}{u} + \frac{1}{v} = \frac{1}{F}.$$

$$\therefore \quad \frac{1}{f_1} + \frac{1}{f_2} = \frac{1}{F} \qquad\qquad \frac{1}{f_1} + \frac{1}{f_2} = \frac{1}{F} \quad......(10.4)$$

The POWER of a lens in DIOPTRE is the reciprocal of its focal length in metre.

The power of thin lenses in contact is thus the algebraic sum of their individual powers in dioptre.

$r = 2f$ for spherical mirrors. Rays striking a spherical concave mirror of radius r parallel to the axis are reflected through the focal point F (Fig. 51). By the laws of reflection, $i = i'$, the normal passing through the centre of curvature C. But $i = i''$ (alternate angles).

$$\therefore \qquad\qquad i' = i'', \qquad \text{or} \qquad HF = FC.$$

As all angles are small, $HF \doteqdot PF$.

$$\therefore \qquad\qquad PF = FC = \tfrac{1}{2}PC.$$

NC	**RP**
$f = -PF, \quad r = -PC.$	$f = PF, \quad r = PC.$
$\therefore \qquad r = 2f$	$\therefore \qquad r = 2f \quad ....(10.5)$

Note that this relationship applies to mirrors only—*not to lenses.*

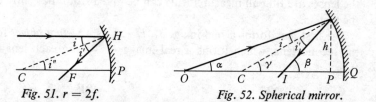

Fig. 51. $r = 2f$.　　　　　Fig. 52. Spherical mirror.

Spherical mirror formula. The concave mirror forms a real image I of the real object O (Fig. 52). As all angles are small, tangents can be approximated to angles in radian, and points P, Q can be considered to be coincident.

Now $\qquad\qquad\qquad\qquad i = i'$.

Also $\qquad\qquad i = \gamma - a \qquad$ and $\qquad i' = \beta - \gamma.$

But $\quad a \simeq \tan a = \dfrac{h}{PO}, \quad \beta \simeq \tan \beta = \dfrac{h}{PI}, \quad \gamma \simeq \tan \gamma = \dfrac{h}{PC}.$

Substituting, $\qquad\qquad \gamma - a = \beta - \gamma,$

or $\qquad\qquad\qquad \dfrac{h}{PC} - \dfrac{h}{PO} = \dfrac{h}{PI} - \dfrac{h}{PC}.$

NC	**RP**
$u = -PO, v = -PI,$ $r = -PC.$	$u = PO, v = PI, r = PC.$
$\therefore \qquad \dfrac{1}{v} + \dfrac{1}{u} = \dfrac{2}{r}.$	$\dfrac{1}{u} + \dfrac{1}{v} = \dfrac{2}{r}.$
But $r = 2f,$	But $r = 2f,$
$\therefore \qquad \dfrac{1}{f} = \dfrac{1}{v} + \dfrac{1}{u} = \dfrac{2}{r}$	$\therefore \qquad \dfrac{1}{f} = \dfrac{1}{u} + \dfrac{1}{v} = \dfrac{2}{r} \quad\ldots\ldots(10.6)$

11. Optical Instruments

Linear and angular magnifications. In these notes, magnifications are considered to be always positive.

LINEAR MAGNIFICATION (or 'Magnification'), m, $= \dfrac{\text{Size of image}}{\text{Size of object}} = \dfrac{h'}{h}.$

ANGULAR MAGNIFICATION (or 'Magnifying power'), M,

$= \dfrac{\text{Angle subtended at eye by image seen through instrument}}{\text{Angle subtended at eye by object seen without instrument}} = \dfrac{\theta'}{\theta}.$

If either object or final image is at infinity, linear magnification has no meaning, and angular magnification must be used.

Magnifying glass. (1) *Image at near point* (Fig. 53b). Linear magnification

$$m = \frac{h'}{h} = \left|\frac{v}{u}\right| = \left|\frac{D}{u}\right|.$$

Substituting for u from the lens formula, where

NC

$$\frac{1}{u} = \frac{1}{v} - \frac{1}{f} = \frac{1}{-D} - \frac{1}{f},$$

RP

$$\frac{1}{u} = \frac{1}{f} - \frac{1}{v} = \frac{1}{f} - \frac{1}{-D},$$

gives

$$m = \frac{D}{f} + 1 \quad \dots\dots\dots\dots\dots(11.1)$$

Now, $\theta' = h'/D$, assuming the eye is close to the lens; and $\theta = h/D$, since the object, when viewed without the instrument, must be moved back to the near point (Fig. 53a).

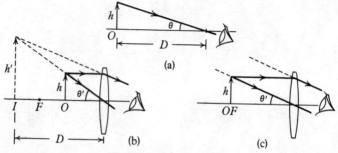

Fig. 53. *Magnifying glass.*

Therefore, angular magnification

$$M = \frac{\theta'}{\theta} = \frac{h'/D}{h/D} = \frac{h'}{h} = m,$$

showing that M is numerically equal to m.

(2) *Image at infinity* (Fig. 53c). Linear magnification has no meaning in this case.

Angular magnification

$$M = \frac{\theta'}{\theta} = \frac{h/f}{h/D} = \frac{D}{f} \quad \dots\dots\dots\dots(11.2)$$

Compound microscope. The object is placed just beyond F of the objective lens, which forms a real, magnified, inverted image at the cross-wires X (Fig. 54). This image is just inside F of the eyepiece, so that the final image formed is virtual and magnified, and— depending on the position of the first image with respect to F—can be located anywhere between the near point and infinity. Fig. 54

48

shows constructional rays drawn for the case of the final image at the near point.

Calculations can usually be accomplished with the aid of a skeleton ray diagram (the *bold* lines in Fig. 54) on which the distances x and y are marked. The relevant lens formulae connecting f, u, v (Eqn. 10.2) and m, u, v (Eqn. 10.3) can then be applied to each lens in turn.

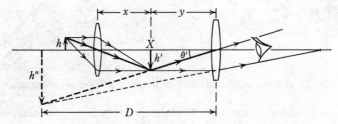

Fig. 54. Compound microscope. Image at near point.

When the final image is at the near point, magnification is equal to magnifying power:

$$M = \frac{\theta'}{\theta} = \frac{h''/D}{h/D} = \frac{h''}{h} = m,$$

and the total magnification is the *product* of the individual magnifications of the component lenses:

$$m = \frac{h''}{h} = \frac{h'}{h} \cdot \frac{h''}{h'} = m_o m_e.$$

Astronomical telescope. The essential difference between the telescope and the microscope is in the objective lens, which in the telescope is of large focal length and diameter, whereas in the microscope it is of small focal length and diameter. The ray diagrams and formation of images are similar, as are the calculations, except, of course, when the telescope is used to view an object at infinity. Four cases can be considered:

(1) Telescope in normal adjustment, i.e. both object and final image at infinity. In this case the focal points of the two lenses coincide (Fig. 55a). In the lower diagram, all three incident rays come from a single point on a distant object *off* the axis (e.g. one edge of the Sun). If the object extends from this point to a point on the axis, then it subtends an angle θ at the eye, as shown, when viewed without the instrument. The final image—at infinity, but

49

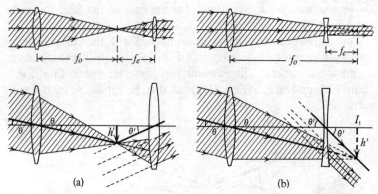

Fig. 55. Astronomical and Galilean telescopes. Normal adjustment.

inverted—subtends the angle θ' at the eye, as shown. For calculation purposes, therefore, only the constructional rays shown as *bold* lines need be drawn. The other rays shown represent a *continuous pencil* of light passing through the instrument.

Thus $\qquad \theta' = h'/f_e \qquad$ and $\qquad \theta = h'/f_o$

and $$M = \frac{\theta'}{\theta} = \frac{h'/f_e}{h'/f_o} = \frac{f_o}{f_e} \quad\dots\dots\dots\dots\dots\dots(11.3)$$

This simple relationship applies only when the telescope is in normal adjustment.

(2) Object at infinity, final image at near point. The eyepiece is moved in slightly, so that the first image falls within the focal length of the eypiece.

(3) Object at a finite distance, final image at infinity. To accommodate a closer object the eyepiece must be moved slightly outwards.

(4) Object at a finite distance, final image at near point. Only in this case, of the four, has linear magnification any meaning. It can be calculated as for a compound microscope—by marking in the distances x and y (Fig. 54) and applying the lens formulae to the two lenses in turn.

Terrestrial telescope. To obtain an erect image, an erecting lens E is included in the astronomical telescope, forming a second image I_2 of the first image I_1, which is of the same size but erected. This second image I_2 now acts as the object for the eyepiece lens (Fig. 56).

The disadvantages of this instrument are that it is increased in length by $4f$ of the erecting lens, the field of view is decreased, and additional aberrations and loss of light are introduced.

Prismatic binoculars. These form a useful alternative to the terrestrial telescope. Erection of the image is accomplished by two 45° prisms at right angles, each of which produces a lateral inversion in one plane only. Objective and eyepiece are similar to those of the telescope (Fig. 56).

Advantages are that a good field of view is retained, while the length is shortened. In addition, a stereoscopic effect is obtained from the use of a separate system for each eye.

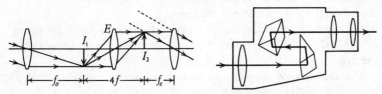

Fig. 56. Terrestrial telescope. Binoculars.

Galileo's telescope. Fig. 55*b* shows the adjustment for object and final image at infinity. The focal points of objective and *concave* eyepiece coincide at the first image I_1. The eyepiece forms an image of I_1 at infinity, which is the correct way up.

The instrument, used for opera glasses, is thus shorter than the astronomical telescope, and gives an erect image; but it has a very small field of view, and the illumination decreases rapidly at points off the axis if the magnification is anything but fairly small.

12. Defects of Vision

Structure of eye. The sclerotic S, a hard, white, nearly opaque, fibrous tissue, encases the eye, merging into the transparent cornea C, a meniscus of radius about 7 mm (Fig. 57). Attached is the coloured iris I, a diaphragm which can vary the size of the pupil P, its central aperture, to suit the intensity of the incident light. The crystalline lens L is held in position by the suspensory ligament SL, and its curvature can be altered by the action of the ciliary muscle CM. The light passes through the transparent aqueous humour A and vitreous humour V, to fall on the retina R. The retina is backed by a dark pigmented membrane, the choroid Ch, which prevents the reflection of scattered light. On the retina the optic nerve O spreads itself, carrying the image impressions to the brain. Lying almost on the principal axis of the lens

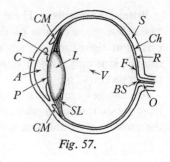

Fig. 57.

is the spot where vision is most acute, the yellow spot, or fovea, F. A blind spot BS occurs at the point where the optic nerve enters the eye.

The yellow spot is made up entirely of 'cones', making it most sensitive to colour and detail. For night vision the part of the retina surrounding it is more useful, since this contains the 'rods', which are sensitive to small quantities of light.

Optical system of eye. The eye and the camera both act as a simple convex lens, forming a real, inverted, diminished image of the object (Fig. 58). The amount of light entering is controlled in both cases by an adjustable diaphragm. The size of this aperture, in the camera, is varied in conjunction with the time of exposure.

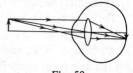

Fig. 58.

The process by which the eye adapts itself to focus objects at different distances is called ACCOMMODATION. Focusing is effected by alterations in the shape of the crystalline lens, which cause it to change its focal length. In the camera it is the image distance, between lens and film, which is altered.

Problems on spectacle lenses

Draw a clear ray diagram using *point* objects and images on the axis, as in Fig. 59. The eye itself does not usually come into the calculation, and may be omitted in the diagram. The function of the spectacle lens is to form a virtual image at a point at which the eye can focus it. The *object,* for the spectacle lens, is always the actual thing, e.g. a book, being observed. The *image* it forms must be at a point at which the unaided eye could focus, and is always virtual.

Do not anticipate the type of lens required by placing a premature

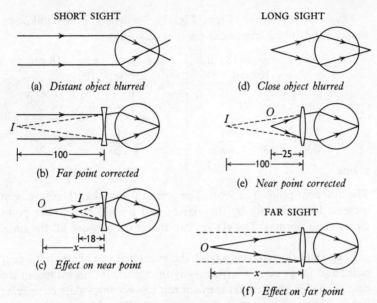

SHORT SIGHT LONG SIGHT

(a) *Distant object blurred* (d) *Close object blurred*

(b) *Far point corrected* (e) *Near point corrected*

FAR SIGHT

(c) *Effect on near point*

(f) *Effect on far point*

Fig. 59. Defects of vision and their correction.

$+$ or $-$ sign before f in the lens formula. The formula must always be applied in the same form, and the sign of f will be indicated in its numerical value at the end of the calculation.

Short sight (myopia). If the eyeball is too long, both near point and far point are closer than normal. Since the far point is at a finite distance, distant objects cannot be seen clearly (Fig. 59a). Correction is by a concave lens, and is best illustrated by an example:

'The far point and near point of a short-sighted man are respectively 100 cm and 18 cm from his eye. What lens is needed to enable distant objects to be seen clearly, and what will be the near point with this lens?'

Correction of far point. Applying the lens formula, where (Fig. 59b)

NC	**RP**
$u = \infty, \quad v = -100 \text{ cm},$	$u = \infty, \quad v = -100 \text{ cm},$
$\dfrac{1}{f} = \dfrac{1}{v} - \dfrac{1}{u}$	$\dfrac{1}{f} = \dfrac{1}{u} + \dfrac{1}{v}$
$\therefore \qquad f = -100 \text{ cm}.$	$f = -100 \text{ cm}.$

The lens required is concave, of power -1 dioptre.

53

Effect on near point. From Fig. 59c, where x is the unknown numerical distance from near point to lens in cm.

$u = -x$ cm, $v = -18$ cm, $f = -100$ cm,	$u = x$ cm, $v = -18$ cm, $f = -100$ cm,
$\dfrac{1}{f} = \dfrac{1}{v} - \dfrac{1}{u}$, or	$\dfrac{1}{f} = \dfrac{1}{u} + \dfrac{1}{v}$, or
$\dfrac{1}{-100} = \dfrac{1}{-18} - \dfrac{1}{-x}$,	$\dfrac{1}{-100} = \dfrac{1}{x} + \dfrac{1}{-18}$,
giving $\quad x = 21.9$	$x = 21.9$

The new near point is 21.9 cm. The person's *range* has therefore been increased enormously by the glasses and he still has a near point closer than normal. The glasses can therefore be worn all the time.

Long sight (hypermetropia). If the eyeball is too short, both near point and far point are further away than normal (this means, in the case of the far point, that the eye at rest can accommodate *converging* light). Close objects cannot be seen clearly (Fig. 59d).

The near point is corrected as in 'far sight' (below) by a convex lens. The effect on the far point in this case is probably merely to bring the far point back to infinity. The glasses can therefore be worn all the time.

Far sight (presbyopia). An ageing person often suffers from a decrease in *range* of accommodation, the near point moving further away but the far point remaining normal. The correction of the near point, by a convex lens, is illustrated by an example:

'A far-sighted man has a near point of 100 cm and his far point is normal. What glasses are necessary to make his near point normal, and what will be his far point when wearing them?'

Correction of near point. From Fig. 59e, taking the normal near point to be at 25 cm,

NC	RP
$u = -25$ cm, $v = -100$ cm,	$u = 25$ cm, $v = -100$ cm,
$\dfrac{1}{f} = \dfrac{1}{v} - \dfrac{1}{u}$	$\dfrac{1}{f} = \dfrac{1}{u} + \dfrac{1}{v}$
$\therefore \qquad f = 33.3$ cm.	$f = 33.3$ cm.

A convex lens is required, of power $+3$ dioptre.

Effect on far point. From Fig. 59*f*, where *x* is the unknown numerical distance from object to lens in cm,

$u = -x$ cm, $v = \infty$,	$u = x$ cm, $v = \infty$,
$f = 33.3$ cm,	$f = 33.3$ cm,
$\dfrac{1}{f} = \dfrac{1}{v} - \dfrac{1}{u}$, or	$\dfrac{1}{f} = \dfrac{1}{u} + \dfrac{1}{v}$, or
$\dfrac{1}{33.3} = \dfrac{1}{\infty} - \dfrac{1}{-x}$,	$\dfrac{1}{33.3} = \dfrac{1}{x} + \dfrac{1}{\infty}$,
giving $\quad x = 33.3$	$x = 33.3$

The range with glasses is thus only 25 cm to 33.3 cm. A far-sighted person will therefore wear either bifocals, or glasses only for reading.

Astigmatism. If the radius of curvature of the cornea in, say, the vertical plane is different from that in the horizontal plane, the eye will have different focal lengths in the two planes. If one plane is in focus the other will be blurred. The appearance to an astigmatic person of a spoked wheel would be as in Fig. 60.

The correcting lens must itself have different radii of curvature in two directions at right angles, and the angle at which the lens is to be fitted into the frame must be specified.

Fig. 60.

13. Geometrical Optics: Experiments

Air cell method for *n* of liquids. The cell consists of two plane, parallel, glass plates separated by a water-tight annulus of paper enclosing an air film.

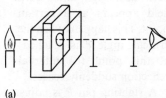

(a)

Fig. 61a. Air cell.

The cell is immersed in the liquid under test (Fig. 61*a*). Monochromatic (sodium) light is observed through the cell along the line of sight of two pins. As the cell is rotated about a vertical axis, two positions are found at which the light is cut off, by internal reflection at the air

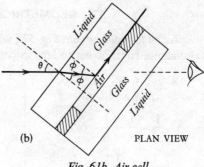

(b)   PLAN VIEW

Fig. 61b. Air cell.

film, as in Fig. 61*b*. The angle 2θ turned through by the cell between these two positions is measured by a pointer and scale. The experiment is repeated to check.

At the cut-off point (Fig. 61*b*) ϕ is the critical angle for the air-glass surface, so $_an_g = \dfrac{1}{\sin \phi}$. Also, $_gn_l = \dfrac{\sin \phi}{\sin \theta}$.

$$\therefore \qquad _an_l = {_an_g} \cdot {_gn_l} = \frac{1}{\sin \phi} \cdot \frac{\sin \phi}{\sin \theta} = \frac{1}{\sin \theta}.$$

Hence $_an_l$ is calculated from θ. The angle θ is seen to be the critical angle for the air-liquid surface, its value being independent of the presence of the intervening glass.

The method is quick and accurate.

Wollaston's method for n of liquids. Only a small quantity of the liquid is needed.

Diffused light from a sodium flame passes into a glass cube or block standing on a drop of the liquid on a dark, matt, horizontal surface (Fig. 62). Internal reflection takes place at the liquid-glass surface, but as the eye is raised vertically, as shown, a line of demarcation between a bright and a dark field appears at a certain point. This happens when the critical angle c is reached: at this point the internal reflection suddenly ceases.

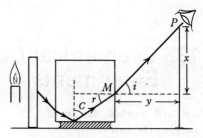

Fig. 62. Wollaston's method.

A sighting pin P is adjusted until the pin, a horizontal mark M on the block, and the line of demarcation, appear in the same straight .

line. The ray through these three points must therefore have been reflected from the liquid-glass face at the critical angle c. The distances x, y are measured, and checked, giving the angle i, where $x/y = \tan i$. Knowing n_g, the unknown n_l is calculated from

$$n_l = \sqrt{n_g^2 - \sin^2 i} \dots\dots\dots\dots\dots(13.1)$$

A convenient method of finding n_g is to perform the experiment first with water, of known refractive index, using Eqn. (13.1) in this case to calculate n_g.

Theory. Eqn. (13.1) is deduced from the following relationships evident from Fig. 62. At the glass-air surface,

$$\sin i = n_g \sin r = n_g \cos c \text{ (since } c = 90° - r\text{).}$$

At the critical angle at the liquid-glass surface (Eqn. 9.4),

$$n_l \sin 90° = n_g \sin c.$$

Eliminating c by squaring and adding these equations,

$$\sin^2 i + n_l^2 = n_g^2.$$

Apparent depth method for n of liquids. A pointer P on the bottom of a tall vessel filled with the liquid forms an image P' as seen from vertically above (Fig. 63). A pointer Q is adjusted vertically until its image Q' by reflection at the liquid surface is coincident with P'.

The method of 'no parallax' is used in this and similar experiments. In the position of no parallax, P' and Q' show no relative motion as the observer moves his head from side to side.

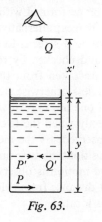

Fig. 63.

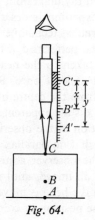

Fig. 64.

For various depths y of the water, x' and y are measured. Each time the adjustment is repeated to check. Then

$$n = \frac{\text{Real depth}}{\text{Apparent depth}} = \frac{y}{x} = \frac{y}{x'} \text{ (since } x = x'\text{).}$$

The pointer Q must be well illuminated from underneath and the vessel should stand on a dark surface. A site free from vibration must be chosen.

Travelling microscope method for n of glass block. The microscope is arranged to travel vertically and is focused on some powder on a hard, smooth surface at A (Fig. 64). The specimen glass block is next placed in position, and the microscope focused on the image of A which is now at B. Thirdly, the microscope is focused on some powder on the top surface C of the block.

From the three vernier readings A', B', C' the differences x and y are found. The experiment is repeated to check. Hence

$$n = \frac{\text{Real depth}}{\text{Apparent depth}} = \frac{AC}{BC} = \frac{A'C'}{B'C'} = \frac{y}{x}.$$

Care must be taken to ensure that the microscope carrier remains firm throughout. It must not, of course, be moved between readings.

Spectrometer method for n of glass prism. (See § 14.)

Lens and mirror constants can be determined by the following standard methods. The theoretical principles underlying these methods should be well understood. In all cases a carefully drawn and labelled ray diagram is essential.

All the experiments described can be performed with pins, using the 'no parallax' technique to locate images; alternatively, most of them can be performed with light-box and screen. Results should always be taken to the nearest mm and checked by repeating the experiment at least once. Readings are taken from the pole of a mirror or from the centre of a lens, unless otherwise stated.

PARALLAX is the apparent motion of one object relative to another due to motion of the observer. It is only when the two objects are coincident that no parallax will be observed. By the method of 'no parallax' the observer adjusts the positions of two images, or an object and an image, until no relative motion is observed between them when the head is moved a short distance from side to side. In this way the position of coincidence is located.

13 . GEOMETRICAL OPTICS: EXPERIMENTS

It is useful to remember, in the experiments which follow, that when an object is *coincident with its own image* the magnification is always unity. The first adjustment in such cases should therefore be to obtain an object and image of approximately the same size.

(1) CONVEX LENS

The four constants n, f, r_1 and r_2 are related by Eqn. (10.2). To determine n we must measure the other three.

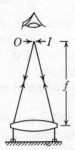

f by plane mirror. Locate position at which object O is coincident with inverted image I formed by reflection from plane mirror (Fig. 65). Measure from this point to top surface of lens; then remove lens and measure instead to mirror surface. If the lens is resting on the mirror, the average of these lengths is the focal length f.

Fig. 65. f of convex lens.

Radii and n by Boys' method. Find f, as above.

Next, place lens on a dark surface. With pin well illuminated from below, find position of coincidence of object O with inverted image I formed by reflection from lower surface of lens (Fig. 66). An erect image, also visible, by reflection from upper surface of lens, is ignored. Measure distance x cm from centre of lens, as before.

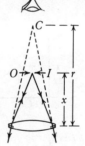

The calculation of the radius r_2 of the lower surface from these readings is now explained. The rays from O are refracted at the top surface, strike the lower surface normally, and are reflected to return along the same path, forming an image coincident with O. Some of the light is not reflected at the lower surface but passes straight through, without deviation, and if the object had been observed by

Fig. 66. Radii of convex lens.

looking upwards through the lens, a virtual image would have been seen at the point C, at which the rays, produced back, meet. *Thus C is the virtual image of O* and the lens formula can be applied to these two points. The point C, however, is also the centre of curvature of the lower surface, and the distance r cm is its radius.

We know f and x, by measurement. The distance r cm is calculated by applying the lens formula to the object O and its virtual image C, where

59

NC	RP
$u = -x$ cm, $v = -r$ cm,	$u = x$ cm, $v = -r$ cm,
$$\frac{1}{f} = \frac{1}{v} - \frac{1}{u},$$	$$\frac{1}{f} = \frac{1}{u} + \frac{1}{v},$$
hence r.	hence r.

The radius r_2 of the lower surface is numerically equal to r. If the lens is biconvex the radius r_1 can be found similarly by turning the lens over. [NC. r_1 is positive and r_2 is negative. RP. r_1 and r_2 are both positive.]

Hence n from Eqn. (10.2).

Other methods for f. Values of u and v for conjugate positions of object and real image can be obtained, and the results dealt with in various ways to obtain f. Examples of standard graphical methods are illustrated in Fig. 67. The student should satisfy himself that the shapes of the curves and information obtainable are shown correctly. Fig. 68 should be studied carefully until the behaviour of a convex lens is well understood.

See also Newton's method and displacement method (pp. 65–6).

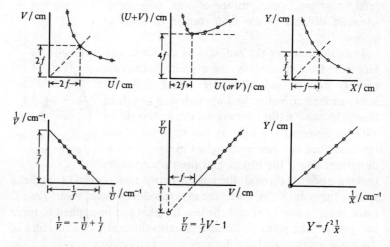

Fig. 67. *Graphs for a convex lens, real objects and images. U, V are the positive distances of O, I from the lens; X, Y are the positive distances of O, I from the focal points (Newton's formula, Eqn. 13.2). The straight line graphs are deduced from the equations given, which are all of the form $y = mx + c$.*

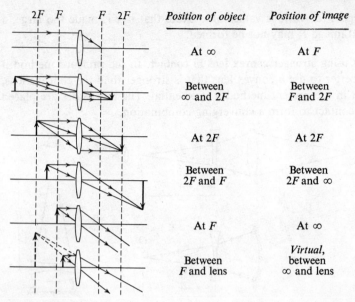

Position of object	Position of image
At ∞	At F
Between ∞ and $2F$	Between F and $2F$
At $2F$	At $2F$
Between $2F$ and F	Between $2F$ and ∞
At F	At ∞
Between F and lens	Virtual, between ∞ and lens

Fig. 68. Object and image positions for a convex lens. Note that the minimum distance between O and I is 4f, and that this happens when U = V = 2f.

(2) CONCAVE LENS

As with the convex lens, n is determined by measuring f, r_1 and r_2 and substituting in Eqn. (10.2).

***f* using auxiliary convex lens.** Locate real image I_1 of object O, formed by convex lens alone, using the 'no parallax' method with a second pin (Fig. 69a). Note position of I_1 along optical bench.

Now interpose concave lens a *short* distance from I_1, towards convex lens, and find new position I_2 of image formed by both lenses. Measure distances x cm and y cm, as shown.

The ray diagram shows that I_2 is the real image of the virtual object I_1 for the concave lens. Therefore,

<table>
<tr><th>NC</th><th>RP</th></tr>
<tr><td>$u = x$ cm, $v = y$ cm,</td><td>$u = -x$ cm, $v = y$ cm,</td></tr>
<tr><td>$$\frac{1}{f} = \frac{1}{v} - \frac{1}{u}$$</td><td>$$\frac{1}{f} = \frac{1}{u} + \frac{1}{v}$$</td></tr>
<tr><td>Hence f.</td><td>Hence f.</td></tr>
</table>

61

Repeat for various values of x. Note that if x is made too large, a real image I_2 may not be formed.

f **using stronger convex lens in contact.** In the previous method it is better to use a convex lens that is stronger than the concave lens, but in the present method it is essential. The two lenses are placed in contact to form a converging combination.

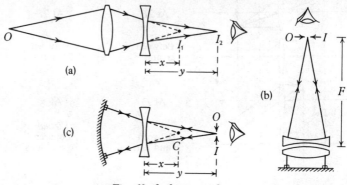

Fig. 69. f of concave lens.

Find F of the combination by the plane mirror method (Fig. 69b), and find f_1 of the convex lens alone by the same method. Calculate f_2 of the concave lens by Eqn. (10.4).

f **using concave mirror.** Find centre of curvature C of concave mirror alone, by coincidence of object and image. From this point C the rays strike the mirror normally and return along the same path (Fig. 69c). Note position of C.

Now interpose lens, and find new position I of coincidence. Measure distances x cm and y cm, as shown. I is thus the real image of the virtual object C for the lens. Compare the ray diagram in Fig. 69a—the focal length f is calculated in the same way.

Repeat experiment with various values of x.

Radii by reflection. The radii of curvature of a biconcave lens are found simply by reflection from the surfaces, which act as concave mirrors (Fig. 70). Coincidence between the well illuminated object O and its image I is obtained at the centre of curvature. [**NC.** r_1 is negative and r_2 is positive. **RP.** r_1 and r_2 are both negative.]

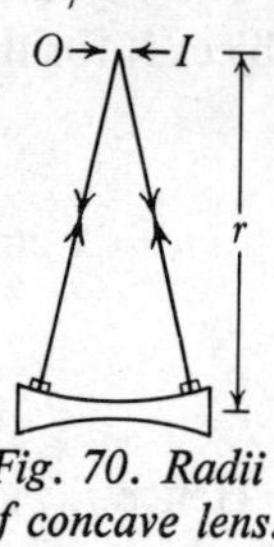

Fig. 70. Radii of concave lens.

13 . GEOMETRICAL OPTICS: EXPERIMENTS

(3) CONCAVE MIRROR

Only one constant has to be determined for a mirror, either f or r, as these are related simply by $r = 2f$ (Eqn. 10.5).

r by coincidence of object and image. Object O and image I coincide at the centre of curvature, since the rays then strike the mirror normally and return along the same path (Fig. 71a). [**NC.** r is negative. **RP.** r is positive.]

f by measurement of u and v. Positions of the real image I for various positions of the object O can be located (Fig. 71b) and f calculated from the mirror formula (Eqn. 10.6).

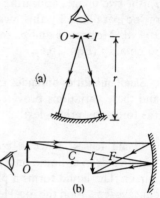

Fig. 71. Concave mirror.

(4) CONVEX MIRROR

r using auxillary convex lens. Place mirror a short distance behind convex lens, and ·adjust object O to obtain coincidence with real image I_1 formed by reflection at mirror (Fig. 72a). In this position the rays strike the mirror normally and return along the same path.

Note position of mirror. Remove mirror, and observe real image I_2 of O formed by lens only, now seen from other side of lens. Locate I_2 by the 'no parallax' method using a second pin. Measure distance r as shown. Clearly I_2 is at the centre of curvature of the mirror and r is numerically equal to its radius. [**NC.** r is positive. **RP.** r is negative.]

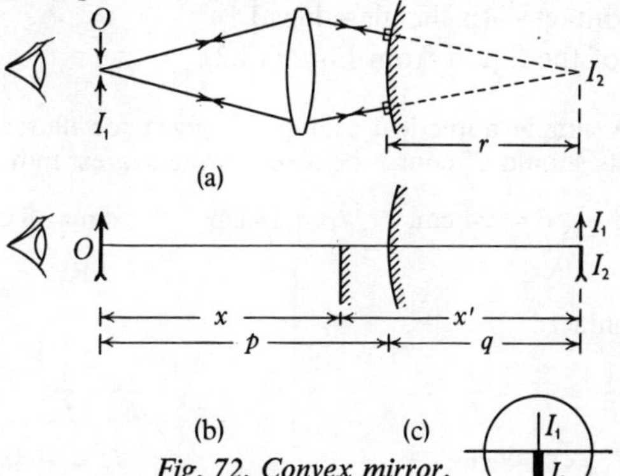

Fig. 72. Convex mirror.

f **using plane mirror.** Place object pin O at about 20 cm from the convex mirror. A virtual, diminished image I_1 is formed (Fig. 72b). Now place plane mirror so as to intercept half of the field of view, and adjust its position until its own image I_2 of O is coincident with I_1 (the two images appearing as shown in Fig. 72c) using the method of 'no parallax'. By this means the position of I_1 is located, since $x = x'$. Measure x and p, and deduce q. Apply the mirror formula to obtain f (Eqn. 10.6).

Small quantities of liquids may be introduced in some experiments, and these variations provide useful additional practice in applying the foregoing principles:

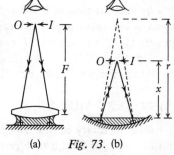

Liquid lenses. Place a convex lens on a drop of liquid on a plane mirror. The liquid forms a plano-concave lens. Find the focal length F of the combination (Fig. 73a), and the focal length f_1 of the convex lens alone, by the plane mirror method. Calculate f_2 of the liquid lens by Eqn. (10.4).

(a)   Fig. 73. (b)

Next, obtain coincidence of object and image formed by reflection from lower surface of convex lens, and measure the distance x cm of this point from the lens (Boys' method, p. 59). Calculate the radius r_2 of the lower surface of the lens. This is numerically equal to the radius r_1 of the upper surface of the liquid lens. [**NC.** r_1 and r_2 are both negative. **RP.** r_2 of the glass lens is positive, but r_1 of the liquid lens is *negative*, since the liquid lens is effectively a lens *in air*, merely in contact with the glass lens.]

Hence n of the liquid from Eqn. (10.2).

EXAMPLE.—A simple numerical example is given for illustration—actual measurements should of course be taken to the nearest mm.

Readings: $F = 24$ cm, $f_1 = 16$ cm, x cm $= 8$ cm.

NC	RP

Lenses in contact:

$$\frac{1}{F} = \frac{1}{f_1} + \frac{1}{f_2} \qquad\qquad \frac{1}{F} = \frac{1}{f_1} + \frac{1}{f_2},$$

$\therefore \qquad f_2 = -48$ cm. $\qquad\qquad f_2 = -48$ cm.

13 . GEOMETRICAL OPTICS: EXPERIMENTS

Boys' method:

$u = -8$ cm, $v = -r$ cm.

$$\frac{1}{f_1} = \frac{1}{v} - \frac{1}{u}, \text{ or } \frac{1}{16} = \frac{1}{-r} + \frac{1}{8}.$$

$\therefore$ $r = 16$,

$\therefore$ r_2 of convex lens $= -16$ cm,
r_1 of liquid lens $= -16$ cm.

n of liquid lens:

$r_1 = -16$ cm, $r_2 = \infty$.

$$\frac{1}{f_2} = (n-1)\left(\frac{1}{r_1} - \frac{1}{r_2}\right),$$

or

$$\frac{1}{-48} = (n-1)\left(\frac{1}{-16} - \frac{1}{\infty}\right).$$

$\therefore$ $n = 1.33$

$u = 8$ cm, $v = -r$ cm.

$$\frac{1}{f_1} = \frac{1}{u} + \frac{1}{v}, \text{ or } \frac{1}{16} = \frac{1}{8} - \frac{1}{r},$$

$r = 16$,

r_2 of convex lens $= 16$ cm,
r_1 of liquid lens $= -16$ cm.

$r_1 = -16$ cm, $r_2 = \infty$.

$$\frac{1}{f_2} = (n-1)\left(\frac{1}{r_1} + \frac{1}{r_2}\right),$$

$$\frac{1}{-48} = (n-1)\left(\frac{1}{-16} + \frac{1}{\infty}\right).$$

$\therefore$ $n = 1.33$.

Liquid on a concave mirror. Find the position of coincidence of object and real image formed by reflection from concave mirror (a) by itself, and (b) with a small quantity of liquid resting on its surface (Fig. 73b). Compare the ray diagram for (b) with Fig. 43b. Thus n of the liquid is obtained simply by adapting Eqn. (9.5):

$$n = \frac{r}{x},$$

ignoring the depth of the water.

Further methods for convex lenses are available, such as the following. If a convex lens is inaccessible, u, v and f cannot be measured directly, since all these quantities involve measuring lengths which terminate at the lens itself. The following methods can be used in these circumstances. Newton's method is also available for a 'thick' lens, or converging combinations of thin lenses.

Newton's method. *Newton's formula* can be derived using the standard ray construction for a real image (Fig. 74). Two pairs of similar triangles are obtained (indicated by shading). The corresponding sides of the similar triangles are proportional:

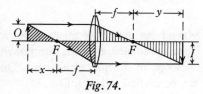

Fig. 74.

65

L.H. pair $\dfrac{I}{O} = \dfrac{f}{x}$,   R.H. pair $\dfrac{I}{O} = \dfrac{y}{f}$.

Equating and multiplying out,

$$f^2 = xy \quad\quad\quad\quad\quad\quad\quad\quad\quad\quad\quad\text{(13.2)}$$

Locate the two focal *points* by the plane mirror method. Then find conjugate object and image positions in the usual way, and measure corresponding values of x and y to the previously determined focal points. Hence f from Newton's formula, Eqn. (13.2).

Displacement method. Place object and image pins, or light-box and screen, a fixed distance apart D, which is *greater* than $4f$ of the convex lens. There will now exist *two* positions of the lens, A and B, in either of which an object at O will produce an image at I (Fig. 75). Keeping OI constant, locate these two positions by moving the lens. Measure OI $(=D)$ and AB $(=d)$. Hence calculate f from Eqn. (13.3).

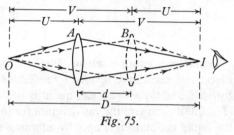

Fig. 75.

Since light rays are reversible, object and image distances are interchangeable. It follows that the two positions A, B are symmetrically placed with respect to O and I, or $OA = BI$, as shown. If the distances U and V are as shown,

$$U = \dfrac{D}{2} - \dfrac{d}{2} \quad\text{and}\quad V = \dfrac{D}{2} + \dfrac{d}{2}.$$

$$\therefore \dfrac{1}{f} = \dfrac{1}{U} + \dfrac{1}{V} = \dfrac{2}{D-d} + \dfrac{2}{D+d} = \dfrac{2(D+d+D-d)}{(D-d)(D+d)} = \dfrac{4D}{D^2-d^2},$$

or

$$f = \dfrac{D^2 - d^2}{4D} \quad\quad\quad\quad\quad\quad\quad\quad\text{(13.3)}$$

Magnifying powers of microscopes and telescopes can be *calculated* from a knowledge of the components, e.g. from Eqn. (11.3) for a telescope in normal adjustment. Direct experimental measurement can be carried out as follows:

M **of microscope.** The microscope is focused on a fine scale divided into, say, hundredths of a millimetre. A cover slip is placed at 45° in front of the eyepiece, so that an image of a millimetre scale, placed 25 cm away, is seen superimposed on the magnified image. By counting divisions the magnification can be determined (Fig. 76).

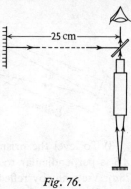

M **of telescope.** If the telescope is focused on a fairly near object, e.g. a brick wall, the magnification can be deduced by a direct comparison of the magnified image

Fig. 76.

seen through the instrument with one eye, and the direct image seen with the other eye.

14. Spectra and the Spectrometer

Description of spectrometer. When correctly adjusted, the vertical slit *A* is at the focal point of the achromatic lens *B* (Fig. 77). The collimator then delivers parallel light to the prism *C*, which refracts each separate colour from the source *S* into a set of parallel rays. Each set is received by the telescope objective *D* at a slightly different angle, so that a pure spectrum is focused on the cross-wires at *E*. The spectrum is observed through the eyepiece *F*.

The telescope, and prism turntable, can be rotated about the same vertical axis. Each can be clamped while the other is rotated and each is provided with a fine-adjustment screw. The angular scale *G* is attached to the telescope, and the vernier scales *H* to the prism table. Rotations can usually be measured to 1 minute of arc. The prism table can be levelled horizontally by three screws *J*.

Preliminary adjustments. (1) Focus eyepiece on cross-wires.

(2) Focus telescope on a distant object. Now, with telescope and collimator in the same straight line, observe the illuminated slit, and focus collimator. The collimator is now focused to deliver parallel light and the telescope to receive it.

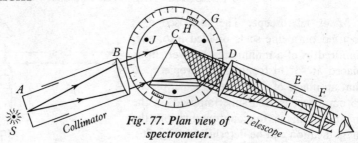

Fig. 77. Plan view of spectrometer.

(3) To level the prism table, place prism as shown, so that face *AB* is perpendicular to the line joining levelling screws *Y, Z* (Fig. 78*a*). View slit by reflection from *AB*, and level table to bring slit centrally in the field of view, in the vertical direction. Repeat, viewing slit by reflection from *AC*, levelling by means of screw *X* only, to avoid disturbing the previous adjustment. The table is now level in *all* planes, having been levelled in two.

Measurement of *n* of prism. An account of this experiment should include the preliminary adjustments to the spectrometer, but without a detailed description of the spectrometer itself. Eqn. (9.7) should be quoted, without proof, unless this is specifically required.

The preliminary adjustments above are first carried out.

Next, to measure the angle *A* of the prism, reduce the slit to a fine line, and, with the table clamped, take vernier readings of the telescope positions when the slit is viewed from the two directions *M*, *N* used in the previous operation (Fig. 78*a*). The angle between these two positions can be shown geometrically to be equal to 2*A*.

To measure the angle of minimum deviation D_{min} for a given spectral line, view the slit by refraction (Fig. 78*b*). Adjust table and telescope so that the required line

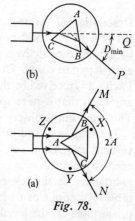

Fig. 78.

appears at the cross-wires when the telescope is in the position of minimum deviation, i.e. when the angle *D* is as small as possible. Take the vernier reading, and, with the table clamped, remove prism and view slit directly through the telescope and collimator in line, i.e. from direction *Q*. Take the vernier reading and deduce the angle D_{min}.

The refractive index n for the given line is calculated from Eqn. (9.7),

$$n = \frac{\sin \frac{1}{2}(A + D_{min})}{\sin \frac{1}{2}A}.$$

Measurement of λ by diffraction grating. (See § 16.)

SPECTRA

Line, band and continuous spectra. An atom, when heated, or bombarded by electrons in a discharge tube, may be 'excited' or even 'ionized'—one or more of its electrons may jump to a state of higher energy, or be knocked right out of the atom. Sooner or later an electron in the atom will revert to a lower energy state, with the emission of the balance of energy as electromagnetic radiation.

Each possible pair of orbits concerned in such an electron jump is associated with a particular amount of energy emitted, and this, in turn, determines the wavelength of the emitted radiation. Thus, if there are n possible jumps, there are n possible amounts of energy emitted, n wavelengths of emission, and consequently n lines in the emission spectrum. It follows that every different kind of atom has its own characteristic line spectrum.

Gaseous atoms, the simplest possible form of matter, thus show line spectra. Gaseous molecules, being more complex systems, have far more possible energy levels available and far more lines in the spectrum. These lines fall into well defined groups, called 'bands', the lines in any one of which are so close as to appear continuous.

Solids and liquids are more complex still, and interactions between adjacent atoms multiply the possibilities to such an extent that the lines merge to form a continuous spectrum.

Thus

> *line spectra:* gaseous atoms
> *band spectra:* gaseous molecules
> *continuous spectra:* solids and liquids.

Emission and absorption spectra. Consider the 'continuous emission' spectrum of a tungsten filament lamp when observed through a sodium flame (Fig. 79). The hot sodium atoms in the flame are capable of absorbing, by 'resonance', exactly the same wavelengths as they can emit. The sodium flame therefore absorbs two visible lines, of wavelengths 0.5890 and 0.5896 μm. This energy is emitted again, at the same wavelengths, by the sodium—but now *in all directions*, so that only a small fraction continues to travel in the direction of the incident beam. The energy received by the spectro-

Fig. 79.

scope is therefore deficient at these wavelengths, and dark lines appear in the spectrum against the continuous background. This is called the 'line absorption' spectrum of sodium, the dark lines appearing in the same places as the bright lines in its emission spectrum. Similarly, gaseous molecules, e.g. in the non-luminous bunsen flame, give 'band absorption' spectra in these circumstances.

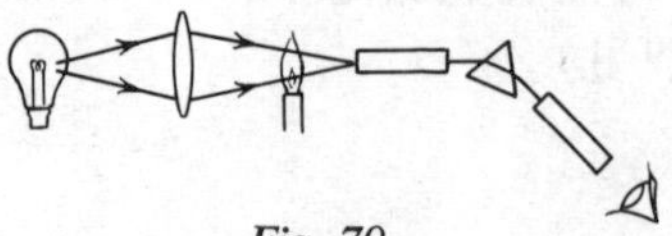

The sun's spectrum. The Fraunhofer dark lines crossing the sun's spectrum are due to the absorption by the cooler gases and vapours in the chromosphere—the outer region of the sun—and by our own atmosphere, of the continuous spectrum emitted by the sun's photosphere—an inner layer. These lines coincide with those of many elements, notably hydrogen.

Line emission

Band emission

Continuous emission

Line absorption

Fig. 80. Examples of types of spectra.

Classification of spectra. A spectrum is thus classified as emission or absorption, and in addition as line, band, or continuous, e.g.

mercury discharge tube: *line emission*
CO_2 discharge tube: *band emission*
filament lamp: *continuous emission*
sun's spectrum: *line absorption.*

15. Photometry

Definitions. Photometry is concerned with the flow of light energy. Three basic quantities are involved—the light energy itself, flowing through space; the intensity of sources emitting the light energy; and the amount of light falling on surfaces, called the illumination. Note that the illumination on a surface refers to the incident light, not the reflected or scattered light.

LUMINOUS FLUX (Φ) is the light energy per unit time emitted from a source or crossing an area in space. *Unit:* lumen.

LUMINOUS INTENSITY (I) of a point source is the luminous flux emitted per unit solid angle in the given direction. *Unit:* candela.

ILLUMINATION (E) on a surface is the luminous flux per unit area falling on the surface. *Unit:* lux.

In order to define the sizes of the units of these quantities we need to define a STANDARD SOURCE. This is a black body radiator at the freezing point of platinum. From this starting point the various units are derived thus:

The CANDELA is the luminous intensity, in a perpendicular direction, of a surface of 1/600 000 square metre of a black body at the temperature of freezing platinum under a pressure of 101 325 N m^{-2}.

The LUMEN is the luminous flux emitted per unit solid angle by a uniform point source of intensity 1 candela.

The LUX is an illumination of 1 lumen per square metre.

Solid angle. The total surface area of a sphere subtends a solid angle 4π at its centre. The solid angle ω subtended at the centre of a sphere of radius r by a surface area A of the sphere is therefore

$$\omega = 4\pi\frac{\text{Area of sphere subtended}}{\text{Total area of sphere}} = 4\pi\frac{A}{4\pi r^2} = \frac{A}{r^2} \quad \text{...}(15.1)$$

Photometry formula. By definition, a point source of intensity I emits $I\omega$ in lumen into solid angle ω. If this flux falls normally on a surface area A the illumination is $I\omega/A$ in lumen per square metre (Fig. 81a).

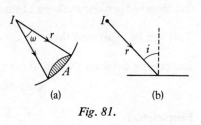

(a) (b)

Fig. 81.

But $\omega = \frac{A}{r^2}$. Therefore the illumination is $\frac{I}{r^2}$.

If the surface is rotated, so that the angle of incidence is i (Fig. 81b),

$$\text{Illumination} = \frac{I\cos i}{r^2} \quad \text{...}(15.2)$$

LIGHT

INVERSE SQUARE LAW IN LIGHT. The illumination on a surface due to a point source varies inversely as the square of the distance from the source.

COSINE LAW OF ILLUMINATION. The illumination on a surface varies as the cosine of the angle of incidence.

Note that Eqn. (15.2) and the inverse square law apply only to radial flow of light from a point source. They do not apply to parts of an optical system in which the rays are *parallel*. In such cases the illumination is undiminished by distance.

Reflection, transmission and absorption. At a boundary between two transparent media part of the incident light is transmitted and part is reflected. Within a medium, light is absorbed, depending on the thickness traversed.

The REFLECTION FACTOR (ρ) of a body is the fraction of the incident flux reflected.

The TRANSMISSION FACTOR (τ) of a body is the fraction of the incident flux transmitted.

The TRANSMISSION COEFFICIENT (k) of a medium is the fraction of the incident flux transmitted by *unit thickness* of the medium.

Let Φ_0 be the incident flux on several layers of a medium, each of unit thickness. Let Φ_1, Φ_2 ... be the flux transmitted by one layer, two layers, etc. Then $\Phi_1 = k\Phi_0$, and $\Phi_2 = k\Phi_1 = k^2\Phi_0$.

In general, where Φ is the flux transmitted by a thickness s,

$$\Phi = k^s\Phi_0 \dots\dots\dots\dots\dots\dots\dots\dots\dots(15.3)$$

The ABSORPTION COEFFICIENT (a) of a medium is the fraction of the incident flux absorbed *per unit thickness* of the medium.

This can be expressed thus, $\quad a = \dfrac{-d\Phi}{\Phi ds}$.

Integrating between Φ_0 and Φ for a thickness s,

$$\Phi = \Phi_0 e^{-as} \dots\dots\dots\dots\dots\dots\dots(15.4)$$

Photometers

The most accurate form of visual photometer is the Lummer-Brodhun, unless the two sources to be compared are of differen colours, in which case the flicker photometer must be used.

Lummer-Brodhun photometer. A, A' are two sides of a diffusing screen, illuminated normally by sources S, S' (Fig. 82). The rays diffused from the screen are internally reflected by prisms B, B' to the cube CC', which consists of two prisms cemented together with Canada balsam at the centre portion only. Light is transmitted at this portion and totally internally reflected elsewhere. Thus the appearance of the field of view changes (Fig. 82a) as the illuminations of A, A' change—A contributing the inner portion, A' the outer. When these illuminations are equal the boundary between the two portions disappears. The remainder of the light is absorbed by the blackened interior of the box D. The eyepiece E is focused on the hypotenuse face of C'.

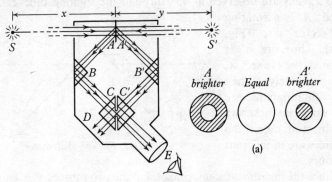

Fig. 82. *Lummer-Brodhun photometer.*

Source S is placed at a suitable distance from A—not too far, as good illumination is best—and S' is moved to the central position within the range in which the boundary is judged invisible. SA (x) and $S'A'$ (y) are measured.

The mean of several readings is taken. The experiment is repeated with the photometer head reversed, to eliminate inequalities in the head itself; and again with S, S' interchanged, to eliminate the effect of stray light. A dark room must be used, and stray light minimized.

Calculation. (1) Comparing two sources. Sources S, S' of intensities I, I' produce equal illuminations on the screens A, A',

$$\therefore \qquad \frac{I}{x^2} = \frac{I'}{y^2}, \quad \text{or} \quad \frac{I}{I'} = \frac{x^2}{y^2}.$$

(2) Finding I if I' is a known standard.

$$I = I'\frac{x^2}{y^2}.$$

(3) Finding the transmission factor (τ) of a glass sheet. S is first matched with S' as above. Then, S remaining in the same position,

73

S' is re-matched (at distance z) with the glass sheet interposed between S' and A'.

Illumination $= \dfrac{I}{x^2} = \dfrac{I'}{y^2} = \tau \dfrac{I'}{z^2}$. Whence $\tau = \dfrac{z^2}{y^2}$.

The mean value of τ is calculated from a number of readings.

Flicker photometer. A, A' are the diffusing screens, coated with magnesium oxide, and illuminated normally by the sources (Fig. 83). The screens are observed at 45° through the sighting tube T. A is fixed; A' is a rotating 90°-sector disc (Fig. 83a). Thus, as A' is rotated, the eye sees A, A' alternately. When the illuminations are the same, no flickering is observed, even if the sources are of different colours.

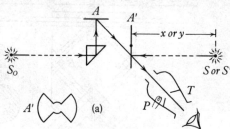

Fig. 83. Flicker photometer.

Since the instrument is unsymmetrical the two sources S, S' cannot be compared directly. Each is compared separately with a source S_0, which remains in a fixed position. If S, S' match S_0 with an absence of flicker at distances x, y, respectively, then $\dfrac{I}{x^2} = \dfrac{I'}{y^2}$.

The best conditions are a small field of view and a bright background—these are provided by lamp P in the sighting tube T. The sources should be fairly close, for high illumination. Speed of rotation of A' should be small. The mean of several readings should be taken. A dark room must be used, and stray light minimized.

Calculations. See Lummer-Brodhun photometer (above).

Photoelectric cells. See 'photoelectric effect', § 45. A photocell can replace the visual photometer if its colour sensitivity curve is similar to that of the eye (Fig. 84). For the alkali metals, notably caesium, maximum photoelectric effect occurs in the visible part of the spectrum, at about 0.5500 μm (= green).

(1) *Barrier-layer type.* A layer of semiconducting material (e.g. cuprous oxide) on the surface of a copper disc is covered with a

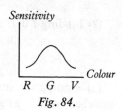

Fig. 84.

very thin transparent metal film (Fig. 85a). Light falling on the top surface of the semi-conductor liberates electrons which pass through the so-called 'barrier layer' into the metal film. The combination acts also as a rectifier preventing the return of the electrons to the semi-conductor. An

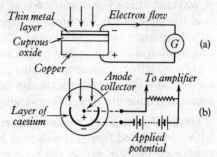

Fig. 85. Photoelectric cells.

external circuit connected as shown will indicate a current, approximately proportional to the intensity of the incident light. No external battery is required for this cell.

(2) *Emission type.* Light falls on the caesium cathode in the vacuum tube, liberating electrons (Fig. 85b). The electrons are drawn to the anode by a considerable p.d. applied externally, and a small current (μA) flows. This is usually amplified by a triode. If the tube contains some argon or helium a larger current is obtained because of the secondary electrons liberated by gas collision.

Calibration of photocell. A known standard source is placed at various distances from the cell and the galvanometer reading noted. Each reading corresponds to a value of illumination which can be calculated from Illumination $= \dfrac{I}{r^2}$. The scale can thus be marked directly in lux.

16. Interference and Diffraction

Properties of waves

For definitions and further discussion of waves, see § 30. For a mathematical treatment of wave motions, see § 33.

Rays and wavefronts. In geometrical optics, light is represented by rays, which show the direction of propagation. When considering light as waves, it is often convenient to draw wavefronts instead.

LIGHT

A WAVEFRONT is a line or surface at all points on which the vibration is in the same phase. For example, ripples on water are wavefronts.

An important fact is that, in an isotropic medium, the direction of propagation is at every point perpendicular to the wavefront. Thus a *plane* wavefront represents parallel rays from an object at infinity; a *spherical* wavefront represents rays diverging radially from a point object at a finite distance, or converging to a point image.

Huygens' principle of secondary waves. Huygens was contemporary with Newton, and they may well have been considered in their day as the champions of the wave, and corpuscular, theories of light, respectively. According to HUYGENS' PRINCIPLE the progression of a wavefront AB (Fig. 86) to a position $A'B'$ an instant later, is considered in this fashion:

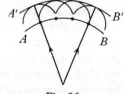

Fig. 86.

(1) every point on the wavefront AB may be regarded as a new source of secondary waves, propagated in all directions, and

(2) the resultant new wavefront $A'B'$ formed is the envelope of these secondary waves.

Diffraction of waves. Huygens' principle accounts for DIFFRACTION —the phenomenon of waves bending round, or spreading out behind, obstacles.

The smaller the obstacle or aperture, the more bending takes place (Fig. 87). This can be demonstrated for water waves by a ripple tank. Appreciable bending occurs only if the obstacle is small enough to be comparable in size to the wavelength. Sound waves always bend round corners because the wavelength is large, e.g. for the note 'middle C' (256 Hz) the wavelength is over 1 metre. But the wavelength of light is of the order of 10^{-7} metre, and light is

Fig. 87.

therefore normally regarded as travelling in straight lines, with obstacles casting geometrical shadows. Light can, however, be diffracted when the obstacle or aperture is very small. Newton did not know this, so he may have tended to favour the corpuscular theory.

Interference of waves. Waves can interfere constructively or destructively with one another, according to the PRINCIPLE OF SUPERPOSITION:

When two waves are superposed, the resultant displacement at any point is the algebraic sum of the displacements of the components.

76

Fig. 88 shows how the resultant wave produced by two component waves of the same amplitude a and wavelength λ depends on their *phase relationship* at the point in question:

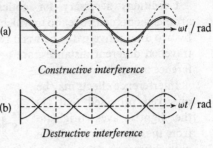

(a)

Constructive interference

(1) Fig. 88a. If the waves are exactly in phase (or 'in step') a wave of amplitude $2a$ results. This is 'constructive interference'.

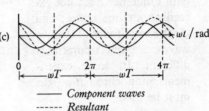

(b)

Destructive interference

(2) Fig. 88b. If they are exactly out of phase (or 'out of step') the resultant wave is zero at every point. This is 'destructive interference'.

(c)

———— Component waves

------ Resultant

Fig. 88. Superposition of waves.

(3) Fig. 88c. If they are in some intermediate phase the amplitude of the resultant is of an intermediate value.

Interference and path difference. If two identical waves, initially in phase at A, travel by different paths ABC, ADC to meet again at C, they will be *in phase* at C if the path difference $(ABC - ADC)$ is an integral number of wavelengths, and constructive interference will occur at C.

If at another point E the path difference $(ABE - ADE)$ is an integral number of wavelengths *plus a half a wavelength*, they will be out of phase at E, and destructive interference will occur at this point.

Thus an 'interference pattern' will be obtained, with alternate maxima and minima of intensity at points in space. The energy, or intensity, of the wave at any point is proportional to the square of the resulting amplitude at that point.

Thus for

Constructive interference:

$$\text{Path difference} = 0, \lambda, 2\lambda, \text{etc.} = n\lambda \quad \dots\dots\dots\dots(16.1)$$

Destructive interference:

$$\text{Path difference} = \tfrac{1}{2}\lambda, \tfrac{3}{2}\lambda, \tfrac{5}{2}\lambda, \text{etc.} = (n + \tfrac{1}{2})\lambda \quad \dots\dots(16.2)$$

where n is any integer.

LIGHT

Conditions necessary for optical interference. Many different arrangements can be set up in which light waves travel simultaneously by two alternative paths between two points. The two waves, having travelled different distances, are no longer synchronized, and interference occurs between them.

Interference effects may be observed provided that (1) the light waves originate from the same source; (2) the path difference does not exceed about one metre; and (3) the amplitudes of the two waves are not appreciably different. In some experiments monochromatic light must be used.

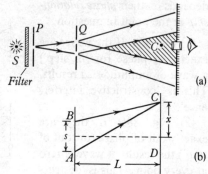

Fig. 89. *Young's experiment.*

Wavelength of light by Young's fringes. Rays from a vertical straight filament pass through the single slit and illuminate the double slit (Fig. 89*a*). Light passing through the double slit is diffracted (or spread out) and interference fringes can be observed in the shaded region of overlap by an eyepiece focused on a point C. P is the vertical, adjustable, single slit and Q the vertical twin slits, ruled about $\frac{1}{2}$ mm apart on aquadag or a fogged photographic plate.

The brightest and sharpest interference pattern is first obtained by trial and error adjustment of (*a*) the width of the single slit, and (*b*) the alignment of the filament, single slit and double slit—the best pattern being obtained when these are exactly parallel.

The micrometer eyepiece is then traversed to measure the distance between a number of fringes, and the distance between adjacent fringes $(x_2 - x_1)$ is calculated. Readings are repeated to check. The distance L from the double slit to the focal plane of the eyepiece is measured with a ruler. The distance s between the twin slits is found by traversing with a microscope focused on the slits, in a separate operation.

The wavelength λ is calculated from Eqn. (16.5). Different filters may be used to find the wavelengths of different colours.

Theory of Young's experiment. A, B are the two slits a distance apart s (Fig. 89*b*). The interference pattern is observed in the plane CD a distance away L. The two rays AC, BC meet at C, a distance x from the 'axis' as shown, and interfere. Their path difference is $(AC - BC)$.

78

Now, by Pythagoras,
$$AC^2 - BC^2 = [L^2 + (x + \tfrac{1}{2}s)^2] - [L^2 + (x - \tfrac{1}{2}s)^2] = 2xs.$$

But   Path difference $(AC - BC) = \dfrac{AC^2 - BC^2}{AC + BC} \simeq \dfrac{2xs}{2L}$

(since $L \gg x$ or s, $AC \simeq BC \simeq L$).

$\therefore$ Path difference $= \dfrac{xs}{L}$ ....(16.3)

Now, bright bands occur at points for which the path difference is $n\lambda$, an integral number of wavelengths (Eqn. 16.1). Therefore in this experiment bright bands occur at distances x from the axis such that

$$n\lambda = \frac{xs}{L} \quad ....(16.4)$$

If the nth bright band occurs at a distance x_1 from the axis and the $(n + 1)$th at a distance x_2, then

$$n\lambda = \frac{x_1 s}{L} \quad \text{and} \quad (n + 1)\lambda = \frac{x_2 s}{L}.$$

By subtraction,

$$\lambda = \frac{(x_2 - x_1)s}{L} ....(16.5)$$

where $(x_2 - x_1)$ is the distance between any two adjacent bands.

Diffraction of Light

A number of simple experiments can be carried out to show qualitatively the diffraction of light. Diffraction patterns are the result of the interference of secondary waves from a wavefront; they can thus be deduced mathematically. The mathematics involved is difficult except in certain simple cases, e.g. a plane diffraction grating receiving parallel light.

Theory of diffraction grating. Parallel rays from the collimator of the spectrometer are incident normally on the grating. When observed through the telescope at an angle θ, a path difference AB is produced between any two corresponding rays from adjacent grating lines (Fig. 90). At an angle θ, where the grating spacing is s,

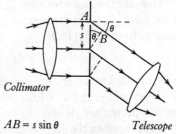

Fig. 90. Diffraction grating.

Path difference $= s \sin \theta$ ....(16.6)

But constructive interference occurs at angles at which the path difference is $n\lambda$, an integral number of wavelengths (Eqn. 16.1). Therefore in this experiment maxima occur at angles θ such that

$$n\lambda = s \sin \theta \quad ....(16.7)$$

Wavelength of light by diffraction grating. The preliminary focusing of the spectrometer is first carried out, as described in § 14.

The experiment is now to set the face of the grating perpendicular to the incident light from the collimator; to ensure that the grating lines are exactly parallel to the slit; and to measure the angles of the diffracted images.

View the slit directly through the telescope. Clamp the table, and rotate the telescope through exactly 90° on the scale. Clamp the telescope. Fix the grating centrally on the table, with the face perpendicular to the line joining levelling screws Y, Z (Fig. 91a). Rotate table until an image of the slit appears on the cross-wires by reflection (Fig. 91b). Use levelling screws Y, Z to bring image vertically into the centre of the field of view. The grating is now at 45° to axis of collimator. Rotate table through exactly 45° to bring grating perpendicular to collimator. Clamp table.

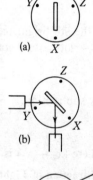

View the first order diffracted image on one side (Fig. 91c) and, using screw X only, so as not to disturb previous adjustment, adjust image to centre of field of view and maximum clarity.

Fig. 91.

This ensures that the grating lines are parallel to the slit.

Measure the angle between the first order diffracted images on the two sides, for a given spectral line (Fig. 91c). This angle is 2θ, where, from Eqn. (16.7),

$$\lambda = s \sin \theta,$$

s being the known grating spacing, the reciprocal of the number of lines per metre, and $n = 1$ for the first order. Hence λ, the wavelength of the spectra line viewed.

Repeat, using the second order spectrum on the two sides. In this case $n = 2$, and

$$2\lambda = s \sin \theta.$$

17. Polarization

Light is a transverse wave motion. It is emitted from a source in a rapid succession of wave-trains, the direction of transverse vibration changing in a random way about 10^9 times per second. This is unpolarized light, the sum effect over a period being as shown (Fig. 92a).

These random transverse vibrations can be resolved into components in any two directions at right angles. If one of these components has been entirely eliminated, the light is said to be *plane polarized*, and to possess a certain 'plane of polarization' (Fig. 92b).

(a)   (b)

Fig. 92. (a) Unpolarized, (b) plane polarized, light approaching observer.

Light falling on the surface of a transparent medium. Most of the light is usually transmitted but part of it is reflected. The reflecting process favours vibrations in one plane against those in another, so that the reflected light becomes partially polarized. The plane of polarization of reflected light is said to be *in the plane of incidence*. At a certain angle of incidence the reflected light is totally polarized, in this plane (see below).

The transmitted light consists of a proportion of unpolarized light plus those polarized components which were not reflected. It is therefore partially polarized in the plane at right angles to the plane of incidence but, unlike the reflected ray, can never be totally polarized.

BREWSTER'S LAW. The reflected light is *totally* plane polarized at a certain angle of incidence i_B, called the POLARIZING ANGLE, such that

$$\tan i_B = n \quad\text{...............................(17.1)}$$

It is seen that this effect occurs when the reflected and refracted rays are perpendicular, or $i + r = 90°$—since in this case $\sin r = \cos i$, and

$$n = \frac{\sin i}{\sin r} = \frac{\sin i}{\cos i} = \tan i.$$

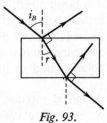

The law is obeyed at both surfaces of a parallel-sided plate (Fig. 93).

Fig. 93.

81

Polarization by reflection. A parallel beam of of monochromatic light, reflected at the polarizing angle (about 57°) from a sheet of glass, is polarized in the plane of incidence. If the reflected beam is intercepted by a similar sheet of glass, so that the angle of incidence is again 57°, and the plane of incidence is perpendicular to that of the first reflection, total extinction of the final beam takes place. The two sheets of glass thus perform the functions, respectively, of *polarizer* and *analyser* (Fig. 94).

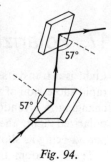

Fig. 94.

Polarization by transmission through a pile of plates. Owing to the large proportion of light transmitted the above method is not very efficient. This disadvantage is overcome by using a pile of plates (Fig. 95), which give a strong beam of transmitted polarized light. At successive reflections (at the polarizing angle) within the plates, more and more of the reflected component is removed from the transmitted beam, which thus becomes more and more polarized itself as it passes through the plates. The beam can be shown to be polarized by using an analyser which, placed at a certain angle, will extinguish the beam.

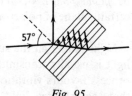

Fig. 95.

Polarization by double refraction. Certain crystals, e.g. calcite and quartz, are 'double refracting'—they transmit *two* refracted beams, polarized at right angles to each other. The beams travel in different directions, and, if they are separated, by a device such as the *Nicol prism*, polarized light becomes available. Tourmaline is double-refracting, but one of the rays is absorbed if the thickness exceeds 1 mm, while the other is almost freely transmitted, providing a polarized beam directly. Polaroid acts in this way. If two sheets of tourmaline or polaroid are placed one on top of the other, and one is rotated in its own plane, successive maxima and minima of intensities of the transmitted light occur every 90°. This is another example of a polarizer and an analyser used in conjunction.

Optical rotation. Some substances exhibit the additional phenomenon of rotating the plane of polarization of light as it passes through them. For quartz, this is 21.72° per mm of its thickness, when sodium light is used. A *polarimeter* consists of polarizer and analyser with the optically active substance placed between them. The angle between the polarizer and the analyser for complete extinction of the transmitted light gives the angle of rotation. The polarimeter is used for the quantitative estimation of sugar solutions, which exhibit this effect; the instrument is then called a *saccharimeter*.

18. Nature and Velocity of Light

Reflection on the particle theory. Newton postulated a repulsive force on the particle when very close to the surface, in a direction perpendicular to the surface (Fig. 96a). Since there is no component of force parallel to the surface the component of velocity in the direction is unchanged. Therefore $v \sin i = v \sin r$, or $i = r$.

This is in accordance with experiment.

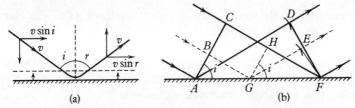

Fig. 96. *Reflection on particle and wave theories.*

Reflection on the wave theory. A plane wavefront ABC is incident on a plane reflecting surface AGF (Fig. 96b). The corresponding points on the reflected wavefront, which by Huygens' principle (p. 74) is the envelope of the secondary waves generated along AGF, are constructed as follows.

Draw an arc, centre A, radius CF, and draw DF to touch this arc at D. With BG, GH drawn paralel to CF, AC, draw an arc, centre G, radius HF, and draw EF to touch this arc at E.

The angle between an incident wavefront and the surface is the angle of incidence, so $\angle CAF = \angle HGF = i$. As the two large triangles ACF, ADF are congruent by construction, $i = \angle AFD$. As the two small triangles GHF, GEF are also congruent, $i = \angle AFE$.

83

Thus $\angle AFD = \angle AFE$ which means that E, and all other points so constructed, must fall on DF. Therefore the reflected wavefront is plane.

Also, since $\angle AFD$ is the angle of reflection r, we have $i = r$. This, again, is in accordance with experiment.

Refraction on the particle theory. Let the particle be entering the more dense medium, with a change of velocity from v_1 to v_2. Newton postulated in this case an attractive force on the particle when approaching very close to the boundary between the media, in a direction perpendicular to it (Fig. 97a). As in the case of reflection there is no component of force parallel to the surface and the component of velocity in this direction is unchanged. Therefore $v_1 \sin i = v_2 \sin r$,

or
$$\frac{\sin i}{\sin r} = \frac{v_2}{v_1}.$$

This is in accordance with the experimental fact that $\sin i/\sin r$ is constant (Snell's law). It claims, further, that $v_2 > v_1$ since by experiment $i > r$. Newton could not measure the velocity of light so was unable to check this point.

Refraction on the wave theory. A plane wavefront ABC is incident on a plane refracting surface AGF (Fig. 97b). Let the velocities in the two media be v_1 and v_2 respectively, and let the second medium be the more dense. The corresponding points on the refracted wavefront, which by Huygens' principle is the envelope of the secondary waves from AGF, are constructed as follows:

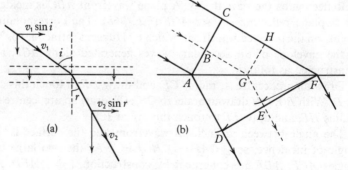

Fig. 97. Refraction on particle and wave theories.

Draw an arc, centre A, radius $\frac{v_2}{v_1}CF$, and draw DF to touch this arc at D. With BG, GH drawn parallel to CF, AC, draw an arc, centre G, radius $\frac{v_2}{v_1}HF$, and draw EF to touch this arc at E.

The angle between an incident wavefront and the surface is the angle of incidence, so $\angle CAF = \angle HGF = i$. But by construction

$$\frac{v_1}{v_2} = \frac{CF}{AD} = \frac{AF \sin i}{AF \sin AFD} \quad \text{and} \quad \frac{v_1}{v_2} = \frac{HF}{GE} = \frac{GF \sin i}{GF \sin AFE}$$

Thus $\angle AFD = \angle AFE$ and it follows that E, and all other intermediate points so constructed, must fall on DF. Therefore the refracted wavefront is plane.

Also, since $\angle AFD$ is the angle of refraction r, we have

$$\frac{\sin i}{\sin r} = \frac{v_1}{v_2}.$$

This is in accordance with the experimental fact that $\sin i/\sin r$ is constant (Snell's law). But it claims, further, that $v_1 > v_2$ since by experiment $i > r$. This is the reverse conclusion to that of the particle theory above.

Light: particle or wave? A characteristic of all wave motions is that the energy is distributed throughout the wave. In the case of particles the energy is essentially in discrete 'packets', with spaces of zero energy in between. It seems therefore that light should be either one or the other—it is difficult to reconcile the two theories.

Briefly, the experimental evidence is as follows:

(1) Reflection slightly favours the wave theory, since Newton had to 'invent' a repulsive force to justify a particle theory.

(2) Refraction is indisputably in favour of the wave theory, as we now know that $v_1 > v_2$ (above) and in the ratio of the refractive index, as predicted:

$$\frac{\sin i}{\sin r} = \frac{v_1}{v_2} = n \qquad \ldots\ldots\ldots\ldots\ldots\ldots(18.1)$$

(3) Interference and diffraction effects can be explained down to the smallest detail by mathematical wave theory, whereas it is difficult to see how particles could destructively interfere.

(4) Polarized light shows different properties in different directions at right angles to its direction of propagation. This may readily be explained by assuming light to be a *transverse* wave motion (p. 136).

LIGHT

The facts of polarization exclude the possibility of light as a *longitudinal* wave as this would have no property which could vary in different transverse directions.

(5) The photoelectric effect (see § 45) and other facts concerning the emission and absorption of light cannot, however, be explained in terms of wave theory—in these situations light behaves unmistakably like particles.

Thus in one situation light behaves as particles, and in another as waves. Electrons likewise show these dual characteristics. The apparent contradiction in this duality of behaviour by both radiation and matter remains one of the puzzles of modern physics and perhaps the only solution lies in a purely mathematical interpretation of events, in which no vivid physical picture is possible. In the meantime, in the words of Sir William Bragg, physicists still 'use the wave theory on Mondays, Wednesdays, and Fridays, and the corpuscular theory on Tuesdays, Thursdays, and Saturdays'.

Measurement of velocity of light

Earlier, astronomical, methods: Römer (1676); Bradley (1727). Terrestrial methods: Fizeau (1849); Foucault (1850); Michelson (1926); Kerr cell methods; radar and other methods.

Michelson's method. Rays of light from the slit A (Fig. 98) fall on one face of the 8-sided mirror B and are reflected on to the large concave mirror C. The distance AC is such that the rays are parallel when reflected from C to a similar mirror D, 35 km distant. The small concave mirror F at the focus of D ensures that the rays are again parallel on reflection back to C. Thence the rays are reflected to another face of B, and an image of the slit A is formed in the microscope E.

The image of A is made to coincide with the cross-wires of E when the mirror B is stationary. B is then rotated, and the speed adjusted until the image again appears on the cross-wires. B now rotates exactly one-eighth of a revolution in the time

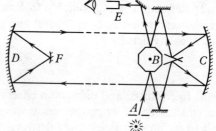

Fig. 98. Michelson's method for velocity of light.

taken for the light to travel the path $BCDFDCB$, a distance d. If the speed of rotation of B in revolution per second, is n, this time is $1/8n$ and the

$$\text{Velocity of light} = \frac{\text{Distance}}{\text{Time}} = 8nd$$

One advantage of the method is that it is a 'null' method, while Foucault's earlier apparatus involved measurement of a finite deflection.

The light is reflected to E for a small fraction only of the total time of revolution, and the intensity received is therefore small, but the intensity was increased by the use of 12- and 16-sided mirrors.

The accuracy of the experiment was limited by the unknown conditions of temperature and pressure between the two stations; in later experiments the light was passed through an iron pipe 1 m wide and over 1 km long, evacuated to a pressure of $\frac{1}{2}$ mmHg.

Velocity in water. This also was measured, and the ratio

$$\frac{\text{Velocity in air}}{\text{Velocity in water}}$$

found to be 1.33, in agreement with the value of the refractive index (Eqn. 18.1).

Part III
Heat

19. Calorimetry

Heat and temperature. Distinguish these two carefully. *Heat* is a form of energy, arising from the motion of the molecules of a substance, and is measured in joule. *Temperature*, in degree Celsius, or kelvin, is a measure of degree of hotness. Hotness is the condition determining which way heat will tend to flow. Heat tends to flow from regions of higher temperature to those of lower, in the same way that water tends to flow down hill, or electricity down a potential gradient. The quantitative definition of temperature is discussed in § 28.

Definitions

The HEAT CAPACITY (C) of a body is the heat required per degree rise in temperature.   *Unit:* $J K^{-1}$.

The SPECIFIC HEAT CAPACITY (c) of a substance is the heat required per unit mass per degree rise in temperature.   *Unit:* $J kg^{-1} K^{-1}$.

The LATENT HEAT (L) of fusion (or vaporization) of a given amount of a material is the heat required to change it from solid to liquid (or liquid to vapour) without change of temperature.   *Unit:* J.

The SPECIFIC LATENT HEAT (l) of a substance is the heat required per unit mass to change its state without change of temperature.   *Unit:* $J kg^{-1}$.

From the definition of specific heat capacity c, the heat Q required to raise mass m from temperature θ_1 to θ_2 is

$$Q = mc(\theta_2 - \theta_1) \quad\text{.......................(19.1)}$$

From the definition of specific latent heat l, the heat Q required to change the state of mass m without change of temperature is

$$Q = ml \quad\text{..............................(19.2)}$$

Examples The following two brief revision examples illustrate the use of the above formulae:

(1) How much heat is required to change 30 gm of ice at $-10°C$ into steam at 100°C?

(2) If 20 gm of ice at 0°C are added to 100 gm of water, initially at 30°C, in a copper calorimeter of mass 50 gm, what final temperature is attained?

Sp. ht. cap. of ice = $2.0 \times 10^3 J kg^{-1} K^{-1}$.
Sp. ht. cap. of water = $4.2 \times 10^3 J kg^{-1} K^{-1}$.
Sp. ht. cap. of copper = $0.4 \times 10^3 J kg^{-1} K^{-1}$.
Sp. lat. ht. of fusion of ice = $336 \times 10^3 J kg^{-1}$.
Sp. lat. ht. of vaporization of water = $2260 \times 10^3 J kg^{-1}$.

WORKING.—(1) Heat required
to heat ice from $-10°C$ to $0°C$	$= 30 \times 10^{-3} \times 2.0 \times 10^3 \times 10$
	$= 600$ J
„ change ice to water at $0°C$	$= 30 \times 10^{-3} \times 336 \times 10^3$
	$= 10\,080$ J
„ heat water from $0°C$ to $100°C$	$= 30 \times 10^{-3} \times 4.2 \times 10^3 \times 100$
	$= 12\,600$ J
„ change water to steam at $100°C$	$= 30 \times 10^{-3} \times 2260 \times 10^3$
	$= 67\,800$ J

$\therefore$ Total heat required $= 91\,080$ J

(2) Heat gained by ice melting $= 20 \times 10^{-3} \times 336 \times 10^3$ J

„ „ „ melted ice warming up $= 20 \times 10^{-3}$
$\times 4.2 \times 10^3 \, (\theta - 0)$ J

„ lost by warm water $= 100 \times 10^{-3}$
$\times 4.2 \times 10^3 \, (30 - \theta)$ J

„ „ „ calorimeter $= 50 \times 10^{-3}$
$\times 0.4 \times 10^3 \, (30 - \theta)$ J

Equating heat gained to heat lost gives $\theta = 12.4°C$.

EXPERIMENTS

The theory of experiments in calorimetry usually involves

$$\text{Heat lost} = \text{Heat gained} \quad ....(19.3)$$

This is valid only if there are no heat losses unaccounted for. In all experiments devices must be employed to eliminate, or reduce, or allow for, heat losses to the surroundings.

Specific heat capacity of solid or liquid by method of mixtures. The solid is placed in a boiling tube which is immersed in boiling water for several minute until the temperature θ_3 is constant. The solid is then quickly transferred to the calorimeter containing cold liquid initially at θ_1. The final maximum temperature θ_2 is noted.

From Eqn. (19.3),

$$m_1c_1(\theta_3 - \theta_2) = m_2c_2(\theta_2 - \theta_1) + m_3c_3(\theta_2 - \theta_1),$$

where m_1c_1, m_2c_2, m_3c_3 refer to solid, liquid and calorimeter, respectively (Fig. 99). Hence either c_1 or c_2, if all else is known.

Heat losses are reduced by the calorimeter jacket. The solid must be heated sufficiently long in the boiling tube to attain a steady temperature; and transferred quickly, avoiding splashing. After transfer, the liquid must be well stirred.

m_1c_1
m_2c_2
m_3c_3

Fig. 99.

HEAT

Specific latent heat of fusion of ice by method of mixtures. Warm water at about 30°C is poured into a previously weighed calorimeter, and weighing is repeated to find the mass of water. The calorimeter is placed in its jacket and the temperature θ_2 is noted. Carefully dried ice is added until the temperature is about 5 degree below room temperature. After stirring, and when all the ice is melted, the final temperature θ_1 is noted. From a final weighing the mass of ice added is found.

From Eqn. (19.3), with temperatures in °C,

$$m_1 l + m_1 c_1(\theta_1 - 0) = m_2 c_2(\theta_2 - \theta_1) + m_3 c_3(\theta_2 - \theta_1),$$

where $m_1 c_1$, $m_2 c_2$, $m_3 c_3$ refer to melted ice, warm water and calorimeter, respectively. Hence l, the only unknown.

Heat losses are almost eliminated if room temperature is halfway between θ_2 and θ_1, in which case heat flows back and forth equally. The ice must be blotted dry immediately before being placed in the calorimeter. The liquid must be well stirred.

Specific latent heat of vaporization of liquid by Henning's method. This electrical method is more accurate than the elementary method of mixtures.

The liquid A under test is kept boiling steadily by an electrical heater B supplying a power VI through the leads C (Fig. 100). The vapour gene-rated passes down the tube D, fitted with a

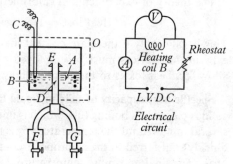

Fig. 100. Henning's apparatus.

splash guard E, and is condensed in F. After a time, the whole appa-ratus becomes warmed to a steady temperature, and a steady state is reached. Now, all the energy supplied electrically is being used to vaporize the liquid. When this state is reached, the vapour is diverted to G for a given time t and the mass m of liquid condensed in this time is drawn off and weighed. Thus energy VIt is supplied electrically (Eqn. 34.7), providing latent heat ml.

Heat losses are almost eliminated by keeping the outer jacket O at the same temperature as the inner container. For this purpose an independent heater is used.

The small residual heat loss (or gain) H, in time t, is allowed for by repeating the experiment with a different wattage, so that a different mass is vaporized. The time t is kept the same, so that H is the same.

Thus $V_1I_1t = m_1l + H$, and $V_2I_2t = m_2l + H$. By subtraction, the unknown H is eliminated:

$$(V_1I_1 - V_2I_2)t = (m_1 - m_2)l \dots\dots\dots\dots(19.4)$$

Hence l.

20. Cooling

Transfer of heat

THERMAL CONDUCTION is the transfer of heat by the action of molecules passing the heat on from one to the next.

The heat flows through the body from places of higher to places of lower temperature.

THERMAL CONVECTION is the transfer of heat by the action of molecules moving and taking the heat with them.

Natural convection currents are set up by the hot body itself in the surrounding fluid, which is heated, expands, becomes less dense, and rises by Archimedes' principle, colder fluid flowing in to take its place. *Forced convection* occurs when a steady stream of fluid is forced past the hot body by some external means.

THERMAL RADIATION is the transfer of heat by electromagnetic waves emitted by one body and absorbed by another.

Radiation can take place through a vacuum. An intervening *gaseous* medium intercepts little of the radiation and is not appreciably heated.

Laws of cooling

NEWTON'S LAW OF COOLING states that, under forced convection, the rate of loss of heat by a hot body is proportional to its excess temperature over that of its surroundings.

This is an empirical law, and holds good for quite large excess temperatures under conditions of forced convection. The relation

93

HEAT

can be applied, as an approximation, to natural convection provided that the temperature differences are small. The heat losses to which the law refers occur, in a very complicated manner, by all three processes of conduction, convection and radiation, though mainly by convection when the temperature differences are fairly small.

A more precise law applicable to natural convection is the FIVE-FOURTHS POWER LAW. This states that,

under natural convection, the rate of loss of heat is proportional to the five-fourths power of the excess temperature of the body over that of its surroundings.

Newton's law mathematically. This can be expressed thus,

$$\frac{dQ}{dt} = -eA\theta \quad ....(20.1)$$

where $-dQ/dt$ is the rate of *loss* of heat (hence minus sign) from a surface of area A at an *excess* temperature θ over that of the surroundings, and e is the emissivity.

The EMISSIVITY (e) of a surface is the heat emitted per unit area per unit time per degree temperature difference between surface and surroundings. *Unit:* $J\,m^{-2}\,s^{-1}\,K^{-1}$.

It depends upon the nature of the surface and the surroundings, and the speed of the draught.

Loss of heat and fall of temperature. The factors influencing *rate of loss of heat* are seen to be

the excess temperature,
the area of the surface,
the nature of the surface,
the nature of the surroundings,
the speed of the draught.

Now, Heat gain = Mass × Sp. ht. capacity × Temperature rise,

or $\qquad \delta Q = mc\delta\theta.$

Similarly, Rate of heat gain = $m \times c \times$ Rate of temperature rise,

or $$\frac{dQ}{dt} = mc\frac{d\theta}{dt} \quad ....(20.2)$$

Combining Eqns. (20.1) and (20.2), where $-dQ/dt$ is the rate of loss of heat and $-d\theta/dt$ is the rate of fall of temperature, we obtain

$$\frac{d\theta}{dt} = -\frac{eA}{mc}\theta \quad ....(20.3)$$

94

Thus the *rate of fall of temperature* is seen to depend upon the above factors and, in addition, upon

the mass of the body,
its specific heat capacity.

Cooling over finite temperature range. The above equations apply, strictly, to instantaneous values. They are sometimes used also for small finite ranges of temperature.

The mathematical student will be able to follow a more rigorous treatment, by integration, for larger ranges. Separating variables and integrating Eqn. (20.3) between excess temperatures θ_0 and θ, for a time of cooling t,

$$\log_e \frac{\theta}{\theta_0} = -\frac{eA}{mc}t,$$

or $\theta = \theta_0 \exp\left(-\frac{eA}{mc}t\right)$ ....(20.4)

and Newton's law results in an exponential cooling curve (Fig. 101).

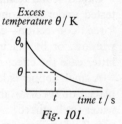

Fig. 101.

Radiation

Thermal radiation is of the same nature as light, differing only in wavelength, being mainly in the infra-red—of larger wavelength than light. The student should be familiar with

(1) demonstrations that the properties of radiation—reflection, refraction, etc.—are similar to those of light;

(2) demonstrations that good absorbers are good emitters;

(3) use of detecting instruments e.g. thermopile; and

(4) pyrometers for measuring high temperatures.

EXPERIMENTS

Determination of melting point. The melting point of a crystalline solid can be found by plotting a cooling curve.

In the simple experiment with naphthalene, a test-tube is half filled with solid naphthalene and heated to 100°C in a water bath. It is then quickly removed from the water and suspended in an empty beaker (which excludes draughts) and allowed to cool. A temperature-time graph is plotted, with readings every half minute, down to about 50°C. The naphthalene is stirred continuously while this is possible.

In the 'ideal' graph obtained (Fig. 102) *AB* represents the cooling liquid, CD the cooling solid. Between B, C the temperature remains constant while the liquid is solidifying—at the melting point, which can thus be accurately determined.

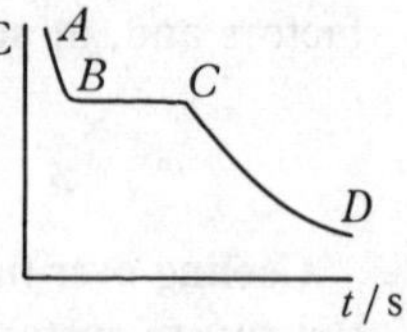

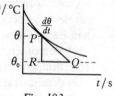

Fig. 102.

The rate of heat loss to the surroundings is the same just before B, just after C, and between B, C—since the excess temperature is the same in all cases. Before B, and after C, however, this heat lost is being provided at a cost of a fall in temperature; whereas between B, C the heat is being provided at a cost of a change of state, involving latent heat.

Experimental testing of Newton's law of cooling. The law must be tested under the conditions to which it is supposed to apply—viz. forced convection—either by placing the cooling body by an open window, or by placing it on a draught produced by a fan. In the latter case the effect of various draughts can also be studied.

A small copper calorimeter, about one-third full of warm water, is placed on a cork slab by an open window. The temperature of the water (well stirred) is taken every minute for an hour. The temperature θ_0 of the draught is also noted. A graph of temperature θ against time t is plotted, and the $\theta = \theta_0$ line drawn in. The gradient $d\theta/dt$ of the curve at several points P is found by drawing the tangent PQ to the curve as accurately as possible at these points (Fig. 103). The corresponding excess temperature $(\theta - \theta_0)$ is represented by PR, and the gradient $d\theta/dt$ by PR/RQ.

Fig. 103.

Now, if Newton's law holds, $\dfrac{d\theta}{dt} \propto (\theta - \theta_0)$, or $\dfrac{PR}{RQ} \propto PR$. Therefore RQ should be constant.

Specific heat capacity of liquid by method of cooling. The method involves allowing water to cool in a copper calorimeter through a given temperature range, then repeating with the same volume of the liquid under test, in the same calorimeter, be-

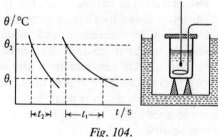

Fig. 104.

tween the same temperatures, in the same external conditions. A graph
of temperature against time is plotted in each case (Fig. 104).

The theory depends on the fact that the cooling surfaces of the
calorimeter are the same whatever liquid it contains. At a given
excess temperature, therefore, both liquids are losing heat at the
same rate. Over an identical temperature range, the *average rate of
heat loss* is the same. This latter deduction can be proved rigorously
by integration and is not restricted to any particular law of cooling.

From Eqn. (20.2), applied to a finite temperature range $(\theta_2 - \theta_1)$,
Average rate of heat loss $=$

$$(m_1c_1 + mc)\frac{\theta_2 - \theta_1}{t_1} = (m_2c_2 + mc)\frac{\theta_2 - \theta_1}{t_2},$$

where m_1c_1, m_2c_2, mc refer to the water, liquid, and calorimeter,
and t_1, t_2 are the times taken for the water and liquid, respectively,
to cool from temperature θ_2 to θ_1.

$$\therefore \qquad \frac{m_1c_1 + mc}{m_2c_2 + mc} = \frac{t_1}{t_2} \qquad \dots\dots\dots\dots\dots(20.7)$$

Hence c_2, where all else is known.

The calorimeter should be in a constant temperature enclosure
(Fig. 104). A lid must be provided to prevent evaporation heat
losses, which would differ for different liquids. Some device for
continuous stirring must be included.

Cooling correction. In those calorimetry experiments in which heat
is supplied at a constant rate *over a period of time*—e.g. the mechanical method (§ 25)—the cooling losses may be sufficient to justify
some simple form of cooling correction.

Let the heat supply begin at A and cease at
B, a time t later (Fig. 105). The temperature
is taken again at C, after a further interval
of time $t/2$. The fall in temperature between
B and C is added to the maximum temperature
recorded at B, to obtain the final temperature
corrected for cooling losses. An example illustrates the reasoning:

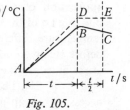

Fig. 105.

Let the measured temperatures at B, C be 23.4°C, 21.8°C, respectively, and the time AB be 8 minute.

∴ Rate of fall of temperature at $B = 1.6$ K in 4 min.

but „ „ „ „ „ „ A = nil.

∴ Average rate of fall between A and $B = 1.6$ K in 8 min.

∴ Corrected final temperature at $B = 23.4 + 1.6 = 25.0°$C.

If no cooling losses had occurred the graph would have followed the 'ideal' curve ADE.

21. Conduction

Heat flow and electricity flow. Among metals, the good conductors of heat are also the good conductors of electricity—this fact indicates that the mechanisms of heat flow and electricity flow are similar. It is instructive to compare the factors on which the flow depends in the two cases:

Rate of electricity flow ∝ Potential difference,
  Cross-sectional area,
  Reciprocal of length of conductor.
Rate of heat flow ∝ Temperature difference,
  Cross-sectional area,
  Reciprocal of length of conductor.

The steady state. An important practical difference between the two, however, is in the time taken to attain a 'steady state'. The application of a p.d. usually results in an almost instantaneous building up of the current to a steady value; but the application of a temperature difference may be followed by a considerable period before steady conditions of heat flow and temperature distribution are attained. In the 'steady state' experiments—e.g. Henning's (§ 19), Searle's (§ 21), Lees' disc (§ 21), Callendar & Barnes' (§ 25)—it is necessary to wait until this state has been attained, i.e. until all temperatures have become *constant*.

In defining thermal conductivity, it is as well to include some phrase such as 'in the steady state'.

Thermal conductivity

The THERMAL CONDUCTIVITY (k) of a material is the rate of heat flow per unit normal cross-sectional area per unit temperature gradient, in the steady state.

Thus
$$\frac{dQ}{dt} = -kA\frac{d\theta}{ds} \quad\text{...........................(21.1)}$$

where dQ/dt is the rate of heat flow down a negative temperature gradient $d\theta/ds$ and through a cross-sectional area A.

The unit of k is $\dfrac{\text{J m}}{\text{s K m}^2}$ or $\text{W m}^{-1}\text{K}^{-1}$.

Temperature distribution along uniform bar

(1) *Bar lagged.* The end X is maintained at a temperature θ_2 (Fig. 106a). Heat flows along the bar and, with perfect lagging, none escapes from the sides. So, in the steady state, the rate of heat flow dQ/dt is the same at all points along the bar. It follows from Eqn. (21.1) that the temperature gradient $d\theta/ds$ is uniform along the bar.

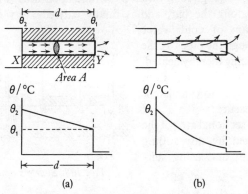

Fig. 106. Temperature distribution along (a) lagged bar, (b) unlagged bar.

If the end Y, a distance away d, attains a temperature θ_1 in the steady state, the temperature gradient $\dfrac{d\theta}{ds} = -\dfrac{\theta_2 - \theta_1}{d}$ and the rate of

heat flow
$$\frac{dQ}{dt} = kA\frac{\theta_2 - \theta_1}{d} \quad\text{........................(21.2)}$$

This is the basic equation for parallel heat flow across plane surfaces.

(2) *Bar not lagged.* As heat escapes from the sides, dQ/dt decreases with distance along the bar. Therefore $d\theta/ds$ decreases also, and a curve is obtained (Fig. 106*b*).

Conduction through composite walls. In Fig. 107, θ_3, θ_2, θ_1 are the temperatures at X, Y, Z; and k_1, k_2 are the thermal conductivities of the materials XY, YZ, which have thicknesses d_1, d_2, respectively.

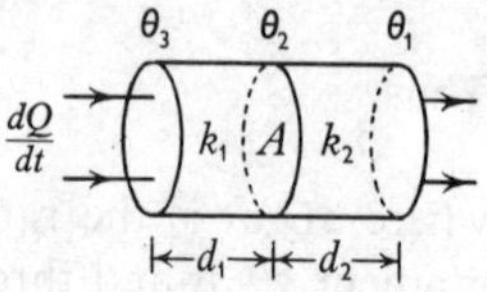

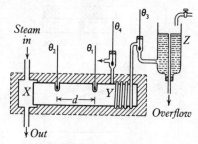

Fig. 107.

The rate of heat flow dQ/dt across each area A is the same,

$$\therefore \quad \frac{dQ}{dt} = k_1 A \frac{\theta_3 - \theta_2}{d_1} = k_2 A \frac{\theta_2 - \theta_1}{d_2} \quad \ldots\ldots\ldots\ldots(21.3)$$

EXPERIMENTS

Searle's method for k of a good conductor. Steam is passed through the steam chest X to maintain one end of a thick bar of the material at about 100°C (Fig. 108). Since the whole apparatus is lagged, the heat flows along the bar, and is finally carried away by a slow, steady stream of cold water flowing through the pipes Y, the cold water being thereby heated.

In the steady state, Eqn. (21.2) applies, so to calculate k we must measure:

A, the cross-sectional area of the bar—obtained by dismantling the apparatus and using suitable calipers;

Fig. 108. Searle's apparatus.

$\dfrac{\theta_2 - \theta_1}{d}$, the uniform temperature gradient along the bar—found from readings of thermometers placed in mercury (for good thermal contact) in holes bored in the specimen a distance apart d; and $\dfrac{dQ}{dt}$, the rate of heat flow through the bar, which equals $\dfrac{mc}{t}(\theta_4 - \theta_3)$,

where m is the mass of cold water passing through the pipes in time t (from weighing and timing), c is the specific heat capacity of water, and θ_3, θ_4 are the steady temperatures of the water entering and leaving the apparatus, respectively.

Heat losses from the sides are reduced by lagging. A 'constant head' device Z must be used to obtain a steady flow of water. All temperatures must be steady before final readings are taken.

Lees' disc method for k of a bad conductor. The above method, suitable for copper ($k = 380$ W m^{-1} K^{-1}), would not be suitable for, say, glass ($k = 0.4$ W m^{-1} K^{-1}) on account of:— the small rate of heat flow; the time taken to reach a steady state; and the heat losses from the sides, which would be comparable with the heat flow down the bar. Clearly dQ/dt must be increased, by reducing d and increasing A. In short, for bad conductors, the specimen must be in the form of a *thin disc*.

In the simple steam-heated form of Lees' disc a thin disc S of the specimen is sandwiched between two thick nickel-plated conducting discs P, Q, the upper of which forms part of a steam chest R (Fig. 109). The apparatus is suspended in a draught-free enclosure and steam is passed. In the steady state the temperatures of two thermometers embedded in holes bored in P, Q will be constant at θ_1 and θ_2 respectively, and

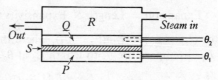

Fig. 109. Lees' disc.

$$\frac{dQ}{dt} = kA\frac{\theta_2 - \theta_1}{d},$$

where dQ/dt is the rate of heat flow through the specimen of thermal conductivity k, cross-sectional area A, and thickness d.

The rate of flow of heat dQ/dt through the specimen is also the rate of loss of heat from the exposed surfaces of P. This is found by a separate experiment in which the steam chest is removed, leaving only the specimen resting on the disc P, suspended as before. P is gently warmed by a bunsen until its temperature is several degrees above its previous temperature θ_1. It is then allowed to cool under the previous external conditions and a cooling curve θ against t is plotted. The rate of fall of temperature $d\theta/dt$ at the temperature θ_1 is obtained from the graph. By Eqn. (20.2),

$$\frac{dQ}{dt} = mc\frac{d\theta}{dt},$$

hence dQ/dt, knowing mass m and specific heat capacity c of disc P.

The dimensions A and d of the specimen are measured, and its thermal conductivity k calculated from the previous equation.

22. Expansion of Solids and Liquids

Solids

The LINEAR EXPANSIVITY of a solid is its fractional increase in length per degree rise in temperature.

The CUBIC EXPANSIVITY is the fractional increase in volume per degree rise in temperature.
Unit (in both cases): K^{-1}.

From the definition of linear expansivity a,

Expansion = Length × Expansivity × Temperature rise ...(22.1)

or $(L_2 - L_1) = L_1 a(\theta_2 - \theta_1)$, where the solid expands from length L_1 at temperature θ_1 to length L_2 at θ_2.

$$\therefore \qquad L_2 = L_1[1 + a(\theta_2 - \theta_1)] \qquad ....(22.2)$$

Similarly $\qquad V_2 = V_1[1 + \gamma(\theta_2 - \theta_1)] \qquad ....(22.3)$

where γ is the cubic expansivity.

Over a limited temperature range, say 0 to 100°C, most solids may be assumed to expand uniformly with temperature. Since the expansivity is usually very small ($\sim 10^{-5}$ K^{-1}) it is unnecessary to specify at which particular temperature the length in Eqn. (22.1) is to be taken—any discrepancy involved will not exceed about one part in 1000.

Relationship between expansivities

Consider a unit cube of the material, heated through 1 K (Fig. 110). By definition, numerical expansion of side = a. Final volume = $(1 + a)^3$ = $1 + 3a + 3a^2 + a^3$.

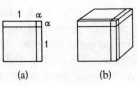

(a)　　　　　　(b)

Fig. 110.

Since $3a \gg 3a^2 \gg a^3$, the increase in volume $\simeq 3a$. But this increase in volume is γ, the cubic expansivity.

$$\therefore \qquad \gamma \simeq 3a \qquad ....(22.4)$$

102

Liquids

The REAL (ABSOLUTE) CUBIC EXPANSIVITY of a liquid is its true fractional increase in volume at 0°C, per degree rise in temperature.

The APPARENT (RELATIVE) CUBIC EXPANSIVITY is the observed fractional increase in volume at 0°C per degree rise in temperature, that is, not taking into account the expansion of the container.
Unit: K^{-1}.

From the definition of the real expansivity a,*

$$V = V_0(1 + a\theta) \quad\quad\quad\quad\text{(22.5)}$$

where the liquid expands from volume V_0 at 0°C to volume V at temperature θ.

But Volume $V = \dfrac{\text{Mass } m}{\text{Density } \rho}$. Therefore $\dfrac{m}{\rho} = \dfrac{m}{\rho_0}(1 + a\theta)$

or $$\rho_0 = \rho(1 + a\theta) \quad\quad\quad\quad\text{(22.6)}$$

The expansivities for liquids (often $\sim 10^{-3}\ K^{-1}$) are appreciably larger than those for solids and it is necessary in the case of liquids to specify 'volume at 0°C', or some other particular temperature, in the definitions and Eqns. (22.5, 6).

Relationship between expansivities

A liquid of real expansivity a is heated in a vessel of *cubic* expansivity g. The resulting apparent expansivity of the liquid is a'. The vessel is initially completely filled and then heated through a temperature rise θ thereby expelling some of the liquid (see weight thermometer, p. 105). During this change the liquid density decreases from ρ_1 to ρ_2, the volume of the vessel increases from V_1 to V_2, and the mass of liquid filling the vessel decreases from m_1 to m_2.

Now $$\rho_1 \simeq \rho_2(1 + a\theta) \quad\text{and}\quad V_2 = V_1(1 + g\theta).$$

$$\therefore \quad \frac{m_1}{m_2} = \frac{\rho_1 V_1}{\rho_2 V_2} \simeq \frac{\rho_1 V_1(1 + a\theta)}{\rho_1 V_1(1 + g\theta)} \simeq 1 + (a - g)\theta.$$

All these approximations are valid in the usual laboratory experiment, where $a\theta$ and $g\theta \ll 1$.

$$\therefore \quad m_1 \simeq m_2[1 + (a - g)\theta] \quad\quad\quad\text{(22.7)}$$

If g is zero, $(a - g)$ becomes the real expansivity a. If g is finite, but is ignored, $(a - g)$ becomes the apparent expansivity a'.

$$\therefore \quad a \simeq a' + g \quad\quad\quad\quad\text{(22.8)}$$

* Since liquids have no linear expansivity, the cubic expansitivity is often denoted by a.

EXPERIMENTS

Comparator method for α of solid. The method aims at a reduction of the principal sources of error—the measurement of the very small expansion, and the attainment of a uniform temperature throughout the specimen.

The specimen A, in the form of a long bar, rests on supports in a double walled water trough BC, the temperature of the outer compartment of which, C, can be controlled thermostatically (Fig. 111). BC is mounted on a truck running on rails. Two lines DD' ruled on the bar are viewed through travelling

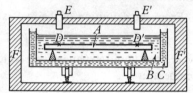

Fig. 111. Comparator method.

microscopes EE' supported on massive stone pillars FF' to prevent movement. Readings of E, E', focused on D, D', are taken with the water bath at various steady temperatures. From the changes in readings the expansion is obtained. The expansivity is calculated from Eqn. (22.1).

To check that no movement of E or E' has occurred between readings a similar standard bar, parallel to the first, and in a similar water bath, is maintained at 0°C in melting ice throughout the experiment. This is wheeled into position under the microscopes at the beginning and end of the experiment. The microscope readings should be the same in both cases.

Balancing columns method for real expansivity of liquid. The L.H limb and tubes A are surrounded by cold water baths at 0°C (Fig. 112). The R.H. limb is surrounded by an oil bath brought up to a steady temperature θ. When the levels in A are steady, heights h_1, h_2, H and temperature θ are measured.

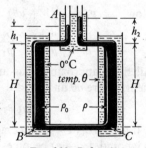

Fig. 112. Balancing columns method.

From Eqn. (2.10), Liquid pressure at $B = h_1\rho_0 g + H\rho_0 g$, and Liquid pressure at $C = h_2\rho_0 g + H\rho g$, where heights and densities are as shown. These two pressures are equal,

$$h_1\rho_0 + H\rho_0 = h_2\rho_0 + H\rho.$$

22 . EXPANSION OF SOLIDS AND LIQUIDS

But $\rho_0 = \rho(1 + a\theta)$ from Eqn. (22.6). Substituting for ρ_0 and re-arranging, where $\qquad H \gg (h_2 - h_1)$,

$$a \simeq \frac{h_2 - h_1}{H\theta}.$$

Hence a. This is the real, not the apparent, expansivity, since the theory depends on hydrostatic pressures, and these are independent of the shape (and expansion) of the containing vessel.

The arrangement at A eliminates certain errors of the simple U-tube version of Dulong & Petit—(1) both tubes at A are at the same temperature, so the surface tension capillary effect is the same for both; (2) there is no exposed column above the constant temperature bath; and (3) the tubes are close together, and comparison of the levels is easier.

Weight thermometer method for apparent expansivity of a liquid.
The weight thermometer, which is a form of density bottle, is weighed empty and dry. It is then filled completely with the liquid under test by alternately heating and cooling it with its open end dipping into a beaker of liquid (Fig. 113). To exclude all air, the liquid should be boiled for some time after partial filling, and then be allowed to refill the bottle as above. The weights of the bottle completely full of liquid at two temperatures are found, and the weights of liquid calculated by subtracting the weight of the bottle.

Fig. 113.

The theory is given on p. 103, from which the equation

$$m_1 \simeq m_2[1 + a'(\theta_2 - \theta_1)] \qquad \ldots\ldots\ldots\ldots\ldots(22.9)$$

is obtained. Hence a' is calculated. If the cubic expansivity g of the glass is known, a can be calculated indirectly, using Eqn. (22.8).

Mathiessen's method. A glass sinker is weighed in air, and totally immersed in the liquid at various temperatures. The *loss* in weight is calculated for each temperature.

By Archimedes' principle, the Loss in weight = Weight of liquid displaced. Therefore the theory as given for the weight thermometer above applies here, where m_1, m_2 are the *losses* in weight at the two temperatures, and the cubic expansivity g is that of the glass sinker.

105

23. The Gas Laws

The gas laws

We can consider the behaviour of a gas under conditions of constant temperature, pressure or volume.

(1) At constant *temperature*:

BOYLE'S LAW. For a fixed mass of any gas at constant temperature, the pressure is inversely proportional to the volume.

Thus $P \propto 1/V$, or $P = k/V$ (k is a constant), or $PV = $ Constant,

or
$$P_1 V_1 = P_2 V_2 \qquad \dots\dots\dots\dots\dots\dots\dots(23.1)$$

(2) Secondly, if we measure, at constant *pressure*, the volumes of various gases at the ice point (0°C) and the steam point (100°C), and divide their respective volume increases over this range by 100, we can calculate their expansivities in 'per degree Celsius' (where the degree Celsius is 1/100 the temperature interval between the ice and steam points). No thermometer is needed for these measurements. Experimentally we find that:

(*a*) the expansivities of all gases, unlike those of solids and liquids, are the same, and

(*b*) the value (to 3 sig. figs.) is 1/273 or 0.00366 per degree Celsius at 0°C.

These two facts can be summarized in Charles' law, as follows:

CHARLES' LAW. Fixed masses of all gases at constant pressure expand by 1/273 of their volumes at 0°C, per degree Celsius rise in temperature.

The CUBIC EXPANSIVITY of a gas at constant pressure is the fractional increase in its volume at 0°C, per degree Celsius rise in temperature.

Its value, as we have seen, is 1/273 per degree Celsius, for all gases.

The ABSOLUTE ZERO is the temperature at which a gas, assumed to continue to contract uniformly with Celsius temperature at the Charles' law rate, would occupy zero volume.

Clearly this temperature would be −273°C.

The ABSOLUTE SCALE OF TEMPERATURE is that scale having its zero at −273°C, the absolute zero, and the size of its degree, or unit temperature interval, the same as that of the Celsius scale.

The thermodynamic, or work, scale of temperature. The concept of absolute zero of temperature would appear, from the foregoing, to be based upon fiction, since no gas could be expected to remain gaseous down to this point, nor to occupy zero volume.

From other considerations, however, based upon the principles of thermodynamics (see further discussion, p. 128), it can be shown that such a zero does in fact exist. Furthermore, it can be shown that these two zeros are identical, and that the two scales of temperature—based respectively on the behaviour of 'ideal' gases, and upon the principles of thermodynamics—are also identical. Since the latter is the more fundamental, we refer to the scale as the thermodynamic, or work, scale of temperature. This has as its unit the kelvin (symbol K). The sizes of the kelvin and Celsius temperature intervals are the same.

It follows that if a given temperature is θ in degree Celsius, and T in kelvin, then

$$T = \theta + 273 \text{ degree} \quad \dots\dots\dots\dots\dots(23.2)$$

Note that in SI notation no distinction is made between a 'temperature' and a 'temperature interval': both are denoted by °C or by K.

The formal, more precise, definition of the kelvin is as follows:

The KELVIN is the unit of thermodynamic temperature. It is the fraction 1/273.16 (*exactly*) of the thermodynamic temperature of the triple point of water.

The triple point of water (273.16 K) differs slightly from the ice point (273.15 K). We could therefore write alternatively that the kelvin is the fraction 1/273.15 of the thermodynamic temperature of the ice point. Eqn. (23.2) would then become, more precisely,

$$T = \theta + 273.15 \text{ degree.}$$

Absolute temperature and the behaviour of gases. Returning to the behaviour of gases, it is noted that the expansivity of gases is larger than that of any liquid, and it is therefore even more necessary to specify, in the definition of expansivity, 'volume at 0°C', since the volume at any other temperature would be appreciably different.

For gases we can now write, using the usual symbols,

$$\frac{V_2}{V_1} = \frac{V_0(1 + a\theta_2)}{V_0(1 + a\theta_1)} = \frac{(1 + \theta_2/273)}{(1 + \theta_1/273)} = \frac{(273 + \theta_2)}{(273 + \theta_1)} = \frac{T_2}{T_1},$$

obtaining

$$\frac{V_2}{V_1} = \frac{T_2}{T_1} \quad \dots\dots\dots\dots\dots\dots\dots(23.3)$$

This equation reiterates a fact assumed in the definition of absolute zero and absolute temperature, that the absolute temperature is proportional to the volume of a Charles' law gas throughout its range. Eqn. (23.3) may thus be taken as a *definition* of absolute temperature T in terms of the behaviour of an 'ideal' gas.

The gas laws (continued)

(3) At constant *volume*, the pressure of any gas, measured at the two fixed points, the ice and steam points respectively, will, as in the above case, show an increase of 1/273 of its pressure at 0°C, per degree Celsius rise in temperature.

The absolute zero of temperature is therefore also the temperature at which a gas, assumed to follow this behaviour throughout its whole range, would exert zero pressure.

The LAW OF PRESSURES. Fixed masses of all gases at constant volume increase in pressure by 1/273 of their pressures at 0°C, per degree Celsius rise in temperature.

This law is a direct consequence of the two previous laws, and does not represent any new experimental fact.

The PRESSURE COEFFICIENT of a gas at constant volume is the fractional increase in its pressure at 0°C, per degree rise in temperature.

By similar reasoning to that used in obtaining Eqn. (23.3),

$$\frac{P_2}{P_1} = \frac{T_2}{T_1} \quad \dots\dots\dots\dots\dots\dots\dots\dots(23.4)$$

As in the case of Eqn. (23.3), this equation defines T in terms of the behaviour of an 'ideal' gas.

Experiments with gases

The gas laws are a statement of experimental facts, and hold good to a high degree of accuracy under ordinary conditions. Deviations from them are discussed in § 29. Simple experiments can be devised to keep one of the three variables P, V, or T constant, while the other two are varied.

(1) **T constant.** The volume V of the fixed mass of air trapped in the closed limb of the J-tube is proportional to l, since the tube is uniform (Fig. 114*a*). The pressure P is varied by moving the R.H. limb vertically, so that the value of h changes. Pressure at X = Pressure at Y, or $P = B + h$, where B is the barometric pressure in mmHg. The temperature is constant at room temperature.

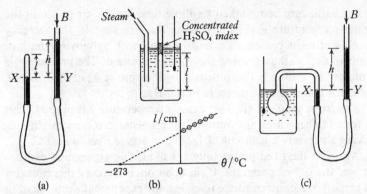

Fig. 114. Experiments with gases at (a) constant temperature, (b) constant pressure, and (c) constant volume.

Results. A series of readings of l against h is taken, and $l(B + h)$ calculated in each case. If this is constant, PV is constant, at constant temperature.

(2) **P constant.** The fixed mass of air is trapped in a capillary tube between the closed end and a conc. H_2SO_4 index which serves to keep the air dry (Fig. 114b). The temperature of the surrounding water bath is varied by bubbling in steam. Time must be allowed for the trapped air to attain the temperature θ of the water, which is then taken with a mercury thermometer. The trapped air expands, and the index moves up, the volume V of the air being proportional to l in the uniform capillary tube. The index adjusts itself to maintain a uniform pressure (atmospheric) in the tube.

Results. A graph of l against temperature θ is plotted. This is found to be—within the limits of experimental error—a straight line. From the slope an expansivity of $1/273$ per degree Celsius at $0°C$ is obtained, and the graph meets the temperature axis at $-273°C$ (Fig. 114b).

Note that Charles' law can be deduced from experiments in which no thermometer is involved. In the above experiment, in which a mercury thermometer is in fact used, for temperatures between the fixed points, an additional conclusion is that the air in the capillary tube appears to expand proportionally to the mercury in the thermometer, within this range of temperatures. (Scales of temperature are discussed in § 28.)

(3) **V constant.** The temperature of the air in the bulb (dried by a small quantity of conc. H_2SO_4) is varied by means of the surrounding

109

water bath, care being taken to allow time for the air to attain the water temperature θ (Fig. 114c). As the air expands, the mercury level is brought back each time to the point X by moving the R.H. limb vertically, thus keeping the volume constant. The pressure P is obtained by noting h. Pressure at X = Pressure at Y, or $P = B + h$, where B is barometric pressure in mmHg.

Results. A graph of $(B + h)$ against temperature θ is plotted. This is found to be, within the limits of experimental error, a straight line, giving a pressure coefficient of 1/273 per degree Celsius at 0°C.

Again, if the fixed points only are used, this experiment demonstrates the law of pressures. If, in addition, a mercury thermometer is employed for intermediate readings, the experiment shows that, in this region, the pressure of the gas increases proportionally to the apparent volume of mercury in the thermometer.

See also constant volume gas thermometer (§ 28).

Ideal gas equation

An IDEAL or PERFECT GAS is one that obeys the gas laws. A real gas deviates slightly from them.

If we write

$$\frac{P_1 V_1}{T_1} = \frac{P_2 V_2}{T_2} \quad \dots\dots\dots\dots\dots\dots\dots(23.5)$$

we see that this expresses the three Eqns. (23.1), (23.3) and (23.4) in a single equation. For example, if T is constant, $T_1 = T_2$, leaving $P_1 V_1 = P_2 V_2$ which is Eqn. (23.1).

These three equations, as we have seen, incorporate the behaviour of an ideal gas, and a definition of an 'ideal gas scale of temperature' T. Eqn. (23.5) is thus the 'ideal gas equation of state', which may be written $PV/T = c$,

or $$PV = cT \quad \dots\dots\dots\dots\dots\dots\dots\dots(23.6)$$

where c is a constant.

The molar gas constant. If the mass of gas is doubled, keeping P and T constant, clearly the volume V is doubled. This means that the constant c is doubled in Eqn. (23.6). Thus c is *proportional to the mass of gas.*

Now, we know that the molar volume of any gas at s.t.p. is $0.0224 \, \text{m}^3 \, \text{mol}^{-1}$. Substituting these values,

$$P = h\rho g = 0.76 \times 13\,600 \times 10 \, \text{N m}^{-2}$$
$$V = 0.0224 \, \text{m}^3 \, \text{mol}^{-1}$$
$$T = 273 \, \text{K},$$

gives $c = \dfrac{PV}{T} = 8.31$ J mol^{-1} K^{-1}. This is the value of c for 1 mole of any gas, and is denoted by R.

The MOLAR GAS CONSTANT (R) for 1 mole of any gas is given by $PV = RT$, where P, V, T are the pressure, volume, and temperature in kelvin, of 1 mole of an ideal gas.

$$R = 8.31 \text{ J mol}^{-1} \text{ K}^{-1}.$$

The MOLE is the amount of substance in a system which contains as many elementary units as there are carbon atoms in 12 g (0.012 kg) of carbon-12.

The AVOGADRO CONSTANT (N_A) is the number of elementary units in 1 mole.

$$N_A = 6.022 \times 10^{23} \text{ mol}^{-1}.$$

The MOLAR MASS (M) is the mass in kg of 1 mole of any substance.

Note that the mole is anomalous in the SI system in that it is based on the gramme rather than the kilogramme. Thus, in SI units, the molar mass of, say, carbon-12 must be given as 0.012 kg mol^{-1}, and not as 12 g mol^{-1}.

The specific gas constant. The molar gas constant R refers to a mass M of the gas (the molar mass in kg mol^{-1}). Similarly we may define a specific gas constant (r) referring to 1 kg of a given gas:

The SPECIFIC GAS CONSTANT (r) for 1 kg of a gas is given by $Pv = rT$, where P, v, T are the pressure, specific volume in m^3 kg^{-1}, and temperature in K, of the given gas.

Since the gas constant is proportional to the mass,

$$\frac{r}{R} = \frac{1 \text{ mole}}{M} \quad\dotfill(23.7)$$

and r is therefore different for different gases.

The specific gas constant is in J kg^{-1} K^{-1}.

The general gas equation. In general, the gas equation referring to mass m of a gas, where M is the molar mass in kg mol^{-1}, is

$$PV = \frac{m}{M}RT \quad\dotfill(23.8)$$

Example. The following example illustrates the importance of specifying 'for a fixed mass' in the gas laws.

Two equal glass bulbs are joined by a narrow tube and the whole is initially filled with air at s.t.p. and sealed. What is the pressure of the air when one of the bulbs is immersed in boiling water and the other in ice?

WORKING.—For the whole system initially at 0°C, $PV = cT$.
  After immersion, for the bulb at 0°C, $P_1V_1 = c_1T_1$.
  After immersion, for the bulb at 100°C, $P_2V_2 = c_2T_2$.
Since total mass of gas is the same before and after, $c = c_1 + c_2$.

$$\therefore \quad \frac{PV}{T} = \frac{P_1V_1}{T_1} + \frac{P_2V_2}{T_2}. \text{ Substituting numerically, } \frac{1.2v}{273} = \frac{P'v}{273} + \frac{P'v}{373},$$

where P' is the unknown pressure in *atmosphere* and v is the volume of each bulb. Solving, $P' = 1.16$ atmosphere, or 878 mmHg.

24. Specific Heats of Gases

Work done on surroundings by a gas expanding. A spring, compressed in the hand, does work against the hand as it is slowly allowed to expand. In the same way a gas, expanding slowly, does work pushing the surrounding atmosphere.*

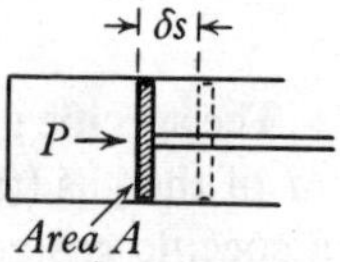

Fig. 115.

Consider a gas at a pressure P pushing back a piston of area A through a distance δs (Fig. 115). Work done on piston by expanding gas = Applied force × Distance moved = $PA\delta s$. But $A\delta s = \delta V$ = Increase in volume of gas.

$$\therefore \qquad\qquad \text{Work done} = P\delta V \text{ in joule} \quad................(24.1)$$

Principal specific heat capacities of a gas

The SPECIFIC HEAT CAPACITY AT CONSTANT VOLUME (c_v) is the heat required per unit mass per degree rise in temperature, at constant volume.

The SPECIFIC HEAT CAPACITY AT CONSTANT PRESSURE (c_p) is the heat required per unit mass per degree rise in temperature, at constant pressure.
Unit: $J\,kg^{-1}\,K^{-1}$.

Difference between c_p and c_v. When a gas is heated at constant volume, the heat supplied is used entirely in raising the temperature. At constant pressure, however, the gas expands, and an additional amount of energy is needed to push back the surrounding atmosphere.

---

* The analogy must not, of course, be taken too far: the energy of the spring is potential, while that of the gas is almost entirely kinetic.

The value of c_p exceeds the value of c_v by the energy involved in the expansion.

Consider an ideal gas heated through a temperature δT at constant pressure P. Total heat required = Mass × Sp. ht. capacity × Rise in temperature = $mc_p\delta T$. Heat required *per unit mass* = $c_p\delta T$. This is made up of $c_v\delta T$ to raise the temperature, together with $P\delta v$ for the expansion (Eqn. 24.1), where δv is the expansion of specific volume v.

$$\therefore \qquad c_p\delta T = c_v\delta T + P\delta v.$$

For unit mass of an ideal gas, $Pv = rT$, where r is the specific gas constant, and v is the specific volume. Differentiating this equation for expansion at constant pressure we obtain $P\delta v = r\delta T$.

Substituting for $P\delta v$ above, $c_p\delta T = c_v\delta T + r\delta T$,

or $$c_p - c_v = r \qquad\dots\dots\dots\dots\dots\dots(24.2)$$

EXPERIMENTS

Joly's steam calorimeter method for c_v.
The two identical copper spheres A, B are exhausted of air and counterpoised on the balance C (Fig. 116). Sphere A is now filled with the gas under test to a pressure of about 20 atmosphere. The counterpoising is repeated, and the mass M of gas admitted is found. The temperature θ_1 of the apparatus, when steady, is measured. Steam is now admitted into the chamber D for about 5 minute through the inlet E, the excess steam escaping through F. The

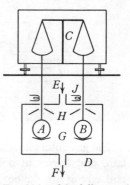

Fig. 116. Joly's differential steam calorimeter for c_v.

whole of the inside of the chamber is thus brought up to the temperature θ_2 of the steam, the steam condensing on the spheres, etc., and giving out its latent heat. The mass of water condensed on sphere A exceeds that condensed on sphere B by the amount required to raise the temperature of the *gas* in A from θ_1 to θ_2. This excess mass m is found by counterpoising again.

Heat required to heat gas = $Mc_v(\theta_2 - \theta_1)$. Heat supplied for this purpose by condensed steam = ml, where l is the specific latent heat of vaporization of water. Thus $ml = Mc_v(\theta_2 - \theta_1)$. Hence c_v, the specific heat capacity at constant volume.

The trays G catch moisture condensed on the spheres, and the shields H prevent moisture dripping on to the spheres from the roof of the chamber. The small electrical heaters J evaporate any drops of water tending to form in the holes, obstructing the suspension wires and free movement of the balance.

The gas is not quite at constant volume, as the sphere expands on account of the rise in temperature and the increasing pressure of the gas. A correction can be calculated.

The theory assumes the spheres to be of equal heat capacity. This cannot be quite true, but the error is eliminated by a repetition of the experiment with the gas in sphere B. The mean of the two results is taken.

Continuous flow method for c_v. The method is based on that of Callendar & Barnes for liquids (§ 25). The same continuous flow method is used, in this case for a gas flowing round the system under conditions of constant pressure (Fig. 117).

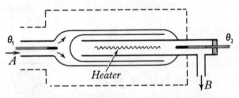

Fig. 117. Continuous flow method for c_p.

The gas enters at A and passes back and forth. It is heated by the coil, and leaves at a higher temperature at B. A simple heating coil circuit would be as in Fig. 100. Temperatures θ_1 and θ_2 are measured. The mass of gas m flowing in a time t is calculated from the density, volume and pressure of gas in the gas reservoir. Thus $VIt = mc_p(\theta_2 - \theta_1)$. Hence c_p.

Heat losses are reduced by a vacuum jacket (not shown). They are further reduced by the arrangement of heat flow—the gas heated by the coil is surrounded by further jackets of incoming gas so that the heat, instead of leaking away, is used again. The whole system is surrounded by a constant temperature water bath.

25. Nature of Heat

Caloric theory and its overthrow. According to this theory heat was an indestructible, weightless, self-repellent fluid, called 'caloric'. The experimental grounds for this theory were reasonable enough—heat obviously 'flows', the hot body losing exactly the amount of heat gained by the cold body. Thermal expansion, it was claimed, was due to the additional room taken up by the caloric flowing into the body.

The theory became untenable when Rumford produced *inexhaustible* supplies of heat from the boring of a cannon barrel. Since it was solely the motion of the borer which produced the heat, the question was then asked: was heat itself therefore motion—the motion, or kinetic energy, of the molecules themselves?

Heat as a form of energy. The new theory was put on a quantitative basis by Joule. The reasoning was as follows. Energy is conserved when converted from one form to another. If heat is a form of energy it, too, must follow this law when converted to or from some other form, no matter by what means the transformation is accomplished. Since heat was, at that time, measured in 'calorie', and other forms of energy in 'joule', such reasoning carried the implication that a constant numerical relationship must exist between the calorie and the joule. By a series of careful and varied experiments Joule established the constancy of this conversion factor, thus showing that heat was a form of energy.

In SI units, we no longer use the calorie. If we wish to repeat Joule's experiments, our aim will now be to show the constancy of the specific heat capacities of substances when heat is supplied in various ways.

EXPERIMENTS

Mechanical methods. Mechanical methods are based on the conversion of mechanical energy to heat by friction. The following method is a development by Rowland (1879) of Joule's original experiment.

A calorimeter *A*, fitted with vanes *B*, was suspended by the torsion

115

wire C (Fig. 118). The paddles D could be rotated by a steam engine connected via the belt and pulley E. The consequent frictional torque on the calorimeter, filled with water, was exactly counteracted by a torque produced by masses M attached to the drum H (Fig. 118a). By this device the applied torque T could be determined.

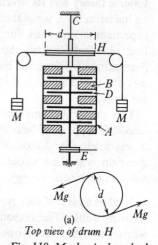

The calorimeter was filled with water and the paddles rotated at a constant speed. The masses were immediately adjusted to oppose the frictional torque exactly, and so bring the calorimeter back to its equilibrium position. The rise of temperature $(\theta_2 - \theta_1)$ of the water, after N revolution of the paddle, was measured with an accurate thermometer.

Work put in $= T\theta$, where T is the torque applied through a total angular rotation θ in radian. The applied torque is equal to the opposing torque Mgd, where d is the diameter of the drum. The angular rotation is $2\pi N$ in radian.

(a)
Top view of drum H
Fig. 118. Mechanical method.

Heat obtained $= C(\theta_2 - \theta_1)$, where C is the heat capacity of the apparatus in $J\,K^{-1}$.

Equating Work put in to Heat obtained, we get

$$C = \frac{Mgd \,.\, 2\pi N}{(\theta_2 - \theta_1)}.$$

Heat losses are reduced by designing the apparatus so that the temperature rises quickly. The longer the experiment takes, for a given temperature rise, the greater the heat losses will be. Heat losses, however, will occur, and must be allowed for by a cooling correction. For a simple method of correction, see § 20.

Callendar & Barnes' electrical method. This continuous flow method was used for investigating variations in specific heat capacity with temperature; also for c_p of gases (§ 24).

Water passes in at A and out at B at a steady rate of mass m in time t, which is measured by weighing and timing (Fig. 119). The water is heated by a coil providing power VI. A simple circuit is

116

shown in Fig. 100. The temperatures θ_1 and θ_2 are measured with thermometers. After some time a steady state is reached and the temperatures remain constant. In this state, almost all the energy supplied

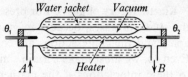

Fig. 119. Callender & Barnes' method.

electrically, VIt in the time t, is being used to heat the water passing in this time. This water absorbs heat $mc(\theta_2 - \theta_2)$, where c is the specific heat capacity of the water.

Heat losses are reduced by the vacuum jacket, and by making $(\theta_2 - \theta_1)$ fairly small. Residual losses are allowed for by a repetition of the experiment with a different rate of flow, the wattage being adjusted by trial and error to give the same temperatures as before. In the same time t, the heat losses H are thus the same in both experiments, since they depend only on t, θ_1 and θ_2, and the temperature of the outer jacket, which is kept constant.

Thus $V_1I_1t = m_1c(\theta_2 - \theta_1) + H$, and $V_2I_2t = m_2c(\theta_2 - \theta_1) + H$

By subtraction, H is eliminated,

$$(V_1I_1 - V_2I_2)t = (m_1 - m_2)c(\theta_2 - \theta_1).$$

Hence c.

26. Kinetic Theory of Gases

Kinetic theory of matter. The theory asserts
(1) that all matter is made up of small particles, and
(2) that these particles are in a state of continuous motion.

A large body of evidence supports these assumptions. This is mainly of an indirect nature, phenomena such as Brownian movement and diffusion perhaps providing the most direct evidence. The kinetic theory gives an entirely satisfactory explanation—outlined in this and following Sections—of the gas laws, the three states of matter, change of state, evaporation, and the properties of vapours.

117

Kinetic theory of gases. For present purposes we limit ourselves to the study of gases, because in gases the particles are relatively far apart and their interactions are consequently the simplest possible. Even so, we must make certain simplifying assumptions with regard to these interactions if the mathematical theory is not to become excessively difficult.

Assumptions made. (1) Gases consist of molecules which behave like hard, smooth, perfectly elastic spheres. These molecules are in continuous, random motion in straight lines, except when they collide with one another and with the walls of the containing vessel.

(2) No appreciable attractive forces are exerted between the molecules.

(3) The volume of the molecules themselves is negligible compared with the total volume of the gas.

Ideal and real gases. We define an ideal, or perfect, gas as one that obeys the gas laws (§ 23). It is also one for which the above assumptions are true. In fact, the breakdown of the gas laws in real, or actual, gases can be ascribed to the breakdown of these assumptions.

To what extent are the three assumptions true for actual gases, and under what conditions do they break down?

(1) This assumption is partly true. The molecules are 'perfectly elastic'—meaning that no kinetic energy is lost in the collisions—otherwise the molecules would quickly come to rest.

They are not perfectly 'hard'—collisions take a finite time to occur, as when a soft rubber ball bounces on the pavement.

Neither are they 'smooth spheres'—a diatomic molecule, for example, is more like a dumb-bell, and possesses rotational and vibrational, as well as translational, kinetic energy.

(2) Because of the attractive forces between molecules, real gases also possess internal potential energy.

(3) The volume of an actual gas could never be reduced to zero, on account of the finite volume of the molecules.

It is clear that assumptions (2) and (3) will tend to break down as the molecules become closer together, i.e. when the gas is subjected to high pressures. Experimentally it is just under these conditions that gases do deviate from the gas laws. On the other hand, we can regard any gas *at infinitely low pressure* as a perfect gas.

Proof of $P = \frac{1}{3}\rho C^2$ for an ideal gas. Consider one molecule, mass m, of an ideal gas, oscillating 'x-wise' as shown, between opposite faces of a rectangular box of sides x, y and z, respectively (Fig. 120). The velocity of the molecule is u.

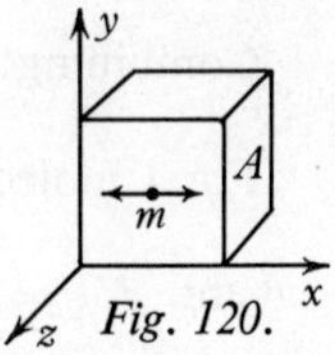

Change of momentum on striking face $A = 2mu$.

Impacts per second on face $A =$

$$\frac{\text{Distance travelled per second}}{\text{Distance between successive impacts}} = \frac{u}{2x}.$$

Fig. 120.

Change of momentum per second at $A = 2mu \cdot \dfrac{u}{2x} = \dfrac{mu^2}{x}$.

But Rate of change of momentum = Force (by Newton's second law of motion), and

$$\text{Pressure} = \frac{\text{Force}}{\text{Area of } A} = \frac{mu^2/x}{yz} = \frac{mu^2}{xyz}.$$

Now, consider N similar molecules, each of mass m, having x-components of velocity of $u_1, u_2, \ldots u_N$.

$$\text{Total pressure on } A = P = \frac{m}{xyz}(u_1{}^2 + u_2{}^2 + \ldots + u_N{}^2)$$

$$= \frac{Nm}{xyz} \cdot \frac{(u_1{}^2 + u_2{}^2 + \ldots + u_N{}^2)}{N} = \frac{Nm}{xyz}\overline{u^2},$$

where $\overline{u^2}$ is the mean square velocity in the x direction. If $n =$ Number of molecules per unit volume $= N/xyz$, then

$$P = \tfrac{1}{3}nmC^2 \quad \ldots\ldots\ldots\ldots\ldots\ldots(26.1)$$

where $C^2 =$ Mean square velocity of the molecules
$= \overline{u^2} + \overline{v^2} + \overline{w^2} = 3\overline{u^2}$ (since $\overline{u^2} = \overline{v^2} = \overline{w^2}$).

$$\therefore \quad\quad\quad\quad P = \tfrac{1}{3}\rho C^2 \ldots\ldots\ldots\ldots\ldots\ldots(26.2)$$

where $\rho = nm =$ Density of the gas.

Interpretation of absolute temperature. Two equations have been deduced for an ideal gas:

$PV = cT$, which defines T on the gas scale of temperature, that is, according to the experimental gas laws (Eqn. 23.6), and

$P = \tfrac{1}{3}\rho C^2$, from the kinetic theory (Eqn. 26.2).

Probably the most fruitful treatment is to fuse these two ideas and see how, together, they produce a third—a more definite interpretation of temperature than was possible previously (§ 19).

For 1 mole, $\qquad PV = RT \qquad$ and $\qquad P = \frac{1}{3}\frac{M}{V}C^2,$

where M and V are the molar mass and volume, respectively.

Combining these, $\qquad RT = \frac{1}{3}MC^2.$

For 1 molecule, $\qquad \dfrac{R}{N_A}T = \frac{1}{3}mC^2,$

where N_A = Avogadro constant (p. 111) = 6.022×10^{23} mol^{-1}
and $\qquad m$ = Mass in kg of 1 molecule.

$\therefore \qquad T = \dfrac{N_A}{3R}mC^2 \quad$ which is proportional to $\frac{1}{2}mC^2.$

Therefore the *absolute temperature on the ideal gas scale is proportional to the average kinetic energy of the molecules of the ideal gas.*

Other deductions from kinetic theory

AVOGADRO'S HYPOTHESIS. Equal volumes of all gases at the same temperature and pressure contain the same number of molecules.

Let two gases be denoted by suffixes 1, 2, respectively, with symbols as above. Then, for

equal volumes: $\qquad n_1, n_2$ both refer to unit volume,
equal pressures: $\qquad P = \frac{1}{3}n_1m_1C_1^2 = \frac{1}{3}n_2m_2C_2^2,$
equal temperatures: $\qquad \frac{1}{2}m_1C_1^2 = \frac{1}{2}m_2C_2^2.$

Dividing, gives $n_1 = n_2$. The simple kinetic theory is therefore in agreement with Avogadro's hypothesis.

DALTON'S LAW OF PARTIAL PRESSURES. The total pressure exerted by a mixture of gases is the sum of the partial pressures, the partial pressure of a component gas being the pressure which it would exert if it alone occupied the whole volume of the containing vessel.

The *total* pressure due to a number of gases would be calculated from the kinetic theory as follows:

$$P = \frac{1}{3}(n_1m_1C_1^2 + n_2m_2C_2^2 + \ldots).$$

The *partial* pressures would be calculated thus:

$$P_1 = \frac{1}{3}n_1m_1C_1^2, \qquad P_2 = \frac{1}{3}n_2m_2C_2^2, \qquad \text{etc.}$$

Therefore $P = P_1 + P_2 + \ldots$ The simple kinetic theory is thus in agreement with Dalton's law of partial pressures.

27. Vapours

Properties of saturated and unsaturated vapours. The apparatus illustrates an investigation into the P-V curve for a vapour (Fig. 121).

When space V is a vacuum, height h represents barometric pressure in mmHg. If a little ether is introduced into this vacuum through the tap, it immediately vaporizes and exerts a pressure, indicated by the decrease in height h. As more ether is introduced the vapour pressure increases; but soon it is noted that the pressure no longer increases, and excess ether liquid

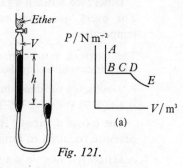

Fig. 121.

collects on the surface of the mercury. The space has become saturated with vapour and can hold no more, and the vapour has attained its 'saturated vapour pressure'—the maximum pressure it can exert at that temperature.

C to B (Fig. 121a). If volume V is now decreased, by raising the mercury levels, the pressure does not increase—as would happen with a gas—but remains constant at the s.v.p., and some of the vapour condenses.

C to D. Similarly, when V is increased, some of the excess liquid vaporizes, to maintain the pressure at its constant value.

A saturated vapour, therefore, does not obey Boyle's law: its pressure is independent of its volume.

D to E. With a further increase in V, if there is only a small amount of excess liquid present, the liquid is presently all vaporized. At this instant the space contains a saturated vapour, but on increasing V beyond D the vapour becomes unsaturated, and its pressure decreases.

An unsaturated vapour can be considered to obey Boyle's law, to a close approximation.

B to A. Finally, when V is decreased until all the space is occupied by liquid, at B, any further attempt to decrease the volume results in a sudden increase in pressure, since a liquid is almost incompressible.

Effect of temperature on s.v.p. The s.v.p. of a pure liquid depends only on its temperature. The increase in s.v.p. with temperature is not uniform, but follows experimental curves such as those shown

121

(Fig. 122). These curves bear no re-
semblance to the *P-T* curve at constant
V of a gas or unsaturated vapour, so
clearly a saturated vapour obeys *none*
of the gas laws.

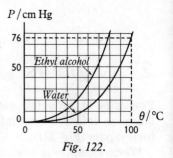

Fig. 122.

Difference between a gas and a vapour.
For every gas there exists a critical
temperature:

The CRITICAL TEMPERATURE of a substance is the temperature above
which it cannot be liquefied, however much it is compressed.

A VAPOUR is a substance in its gaseous state *below* its critical temperature.
If above its critical temperature it is called a GAS.

The closer a vapour is to its critical temperature, the greater the
pressure required to liquefy it (see § 29).

A SATURATED VAPOUR is a vapour in equilibrium with its own liquid.

The (SATURATED) VAPOUR PRESSURE (s.v.p.) of a liquid is the maximum
pressure its vapour can exert at that temperature.

Gas	Unsaturated Vapour	Saturated Vapour
Above critical temperature	Below critical temperature	Below critical temperature
Obeys gas laws	Obeys gas laws approximately	Does not obey gas laws

Kinetic theory applied to vapours. The kinetic theory can be
extended with equal success to explain the properties of vapours.

Effect of temperature on rate of evaporation. Only those molecules
possessing a certain minimum velocity can overcome the attractive
forces of the liquid and escape as vapour. An increase in temperature
means a general step-up in molecular velocities—an increase, therefore in the numbers possessing the required escape velocity. Evaporation will thus proceed faster at higher temperatures.

Why evaporation causes cooling. As only the faster moving molecules can escape, the average velocity of those remaining in the
liquid will fall. Consequently the liquid suffers a fall in temperature.

Latent heat of evaporation. Molecules in the vapour state are
further apart than those in the liquid state. The latent heat represents principally the work done pulling these apart against the
attractive forces. This energy is recovered when the vapour condenses.

Dynamic equilibrium in a closed vessel. Molecules escape from the surface of the liquid at a rate depending only on the temperature. As the space above fills with vapour, so the rate of return of the molecules to the liquid, by chance collision with the surface, increases, until the number returning equals the number

Fig. 123.

escaping, per second. The vapour has now become saturated, and in a state of dynamic equilibrium with the liquid (Fig. 123).

Effect of volume changes on the equilibrium. The equilibrium is temporarily disturbed by the resulting change in vapour density, since the number of molecules re-entering the liquid is affected. This number either rises above (with compression) or falls below (with expansion) the number leaving, so that condensation, or evaporation, takes place until the equilibrium is restored.

Effect of temperature changes on the equilibrium. The number leaving the liquid per second is governed by the temperature. Increasing the temperature therefore increases the number leaving, and so increases the vapour density. The equilibrium is restored at a new level, when the number re-entering has caught up with the number leaving. As the vapour density is now greater, so is the s.v.p.

Mechanism of boiling. *Evaporation* occurs at all temperatures, and takes place only from the exposed surface of the liquid. It is also invisible. *Boiling* occurs at a particular temperature, with bubbles which start from the bottom or sides of the container, wherever heat is applied, and rise through the body of the liquid.

Microscopic cavities exist in the container surfaces which the liquid, because of its surface tension, cannot completely fill. These cavities become filled with saturated vapour. As the temperature rises the s.v.p. increases, but these minute bubbles are prevented from expanding by the opposing pressure P (Fig. 124), which is the sum of the barometric, hydrostatic and surface tension

Fig. 124.

pressures. At the boiling point the s.v.p. just exceeds this opposing pressure, and the bubbles suddenly expand and rise to the surface, where the vapour is liberated.

So we may define the BOILING POINT of a liquid as the temperature at which its s.v.p. is equal to the external pressure. The vapour-pressure/temperature graph (Fig. 122) is thus also the boiling-point/external-pressure graph.

If the surfaces are smooth, the cavities are small, and the opposing surface tension pressure, which depends on curvature, is large. Boiling in bumps then occurs. Pieces of porous material introduced into the vessel provide larger cavities and promote smoother boiling.

Mixture of gas and saturated vapour. A liquid will exert its s.v.p. in any enclosed space above it, irrespective of the presence of any gas pressure in the space. The total pressure P' observed is the sum of the partial pressures—P of the gas, and p of the vapour:

$$P' = P + p \quad\text{............................(27.1)}$$

The apparatus in Fig. 114 can be modified for investigations of the behaviour of mixtures of gas and vapour. In Fig. 114*a* and 114*c* excess liquid can be introduced into the air space. In Fig. 114*b* a liquid index, e.g. water, is substituted for the acid index.

Theory. As an example, the case of the Boyle's law apparatus (Fig. 114*a*) is considered. The student should deduce similar equations for the other two experiments.

The space now contains gas and saturated vapour. Let P', P'' be the observed pressures at the observed volumes V_1, V_2.

Then $\qquad P' = P_1 + p \qquad$ and $\qquad P'' = P_2 + p$,

where P_1, P_2 are the respective gas pressures, and p is the vapour pressure. The vapour pressure is the same in both cases, as it is independent of volume when temperature is constant.

Applying Boyle's law *to the gas only*,

$$P_1 V_1 = P_2 V_2,$$

or $\qquad (P' - p)V_1 = (P'' - p)V_2.$

Refrigerators

The refrigerant, an easily liquefied substance (e.g. ammonia), is contained in a sealed system (Fig. 125). The pump, which may be mechanical or heat-operated transfers vapour from A to B, decreasing the pressure at A and increasing it at B. Liquid evaporates at A due to the decreased pressure, the latent heat required for this being provided by the surroundings which consequently are cooled. At B

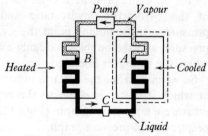

Fig. 125. Refrigerator principle.

liquid condenses, due to the increased pressure, and latent heat is re-sealed: this side of the system is therefore heated.

This transfer of heat from A to B continues so long as the pump operates. Heat is therefore being forced to flow up a temperature gradient, but external energy must be supplied to the pump to make it do so.

The liquid flows back from B to A through C, which limits the rate of flow so that a pressure difference can be built up between the two sides of the system.

EXPERIMENTS FOR S.V.P.

Barometer tube method

(1) *For range* $0°-50°C$ *for water.* A small quantity of the liquid under test is introduced by a bent pipette into the vacuum over mercury in one of the two identical barometer tubes (Fig. 126a). The resulting depression h, measured accurately, gives the s.v.p. The surrounding water jacket is heated electrically or by bubbling in steam, and the s.v.p. measured at various temperatures.

The useful temperature range is limited by the necessity to maintain the whole of the vapour column at a uniform temperature. The vapour pressure will always be that of the coldest part of the tube, since condensation will occur in the coldest part if the prevailing pressure exceeds this value. As the temperature increases in the experiment, it becomes increasingly difficult to maintain uniform conditions over an increasing length of vapour column.

Corrections are needed (1) for pressure of excess liquid resting on

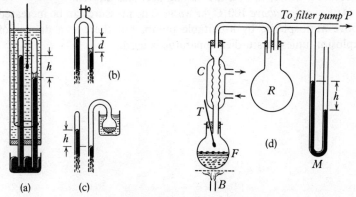

Fig. 126. Methods for measuring s.v.p.

the mercury; (2) for surface tension of this liquid—this can be determined by the apparatus in Fig. 126b; and (3) for reduction of the result from mmHg at the temperature of the experiment to mmHg at 0°C.

(2) *For range below 0°C for water.* The apparatus is modified to Fig. 126c. The bulb contains the solid or liquid under test and is surrounded by a freezing mixture, the temperature of which can be varied by varying the proportions of its constituents. The difference in mercury levels gives the vapour pressure in the coldest part—the bulb itself.

Dynamical method

(3) *For range 50°–100°C for water.* The method is based on the principle that a liquid boils when its s.v.p. is equal to the external pressure.

The external pressure is in this case the pressure inside the system, measured by the mercury manometer *M* in conjunction with a barometer (Fig. 126d). The pressure can be reduced by the filter pump *P*. As the pressure is reduced step by step the liquid boils at lower temperatures, recorded by thermometer *T*. The bunsen *B* keeps the liquid boiling throughout the experiment. The vapour is condensed by the reflux condenser *C* as soon as it is formed, and returns to the flask *F*. The reservoir *R* damps out small oscillations in pressure due to uneven boiling and allows gradual and controlled changes in pressure to be effected more easily.

The range of temperatures is limited by the condenser, which becomes inefficient as the vapour temperature approaches that of the circulating water, so that the boiling becomes uneven.

(4) *For range above 100°C for water.* The pressure can be increased above atmospheric by a suitable pump, but there is a danger of explosion unless a sturdier apparatus is used.

28. Scales of Temperature

The problem. Suppose we have two thermometers A, B containing different liquids. We place both in melting ice, and in steam, and mark the 0°C and 100°C positions on each stem accordingly. In each case we then divide up the length between these two marks into 100 equal divisions, and call these intervals 'degrees Celsius'. Both thermometers are then placed in the same water bath, but we find that A reads 60°C, and B 61°C. Have we any grounds for supposing that both should have read the same temperature? And if not, which is right?

The answer is, of course, that with any thermometer we are not in fact measuring 'temperature' in an absolute sense, but merely some property (in this case the volume of a liquid) which changes with temperature. A 'water thermometer', for example, would, on this basis, show an initial *fall* in temperature from 0°C to 4°C, as water contracts over this range. Neither of the two thermometers A, B is 'right'—each has its *own* temperature scale. For accurate work, therefore, we must define our scales of temperature more clearly.

Establishing a scale of temperature. (1) Choose some property that increases continuously with temperature—e.g. volume of a liquid, pressure of a gas at constant volume, resistance of a platinum wire, e.m.f. of a thermocouple.

(2) Mark the positions of two 'fixed points' on the thermometer scale. Fixed points are temperatures chosen for their ability to be accurately defined, and easily reproduced. For ordinary temperatures we use the ICE POINT (the temperature of pure melting ice at standard pressure) and the STEAM POINT (the temperature of the steam just above pure water boiling at standard pressure).

(3) Divide the scale between these two points into 100 equal parts. We can extend the scale in either direction by further subdivisions of the same size.

Our particular scale of temperature is now defined—e.g. as °C on the 'mercury in glass' scale, or °C on the 'platinum resistance' scale—but our thermometer will record the same temperature as any other *only at the two fixed points*.

127

Definition of a scale of temperature in degree Celsius. A temperature θ in °C may be defined by

$$\frac{\theta}{100} = \frac{X_\theta - X_i}{X_s - X_i} \qquad\qquad\qquad\text{......................(28.1)}$$

where X_θ, X_s, X_i are the values of the property at the given temperature, the steam point, and the ice point, respectively.

Eqn. (28.1) can be deduced from the graph (Fig. 127) since, by similar triangles, $\dfrac{AB}{AC} = \dfrac{BD}{CE}$, giving the equation required.

Making ADE a straight line is equivalent to dividing the scale into 100 *equal* parts, or to assuming that the property increases uniformly with temperature. The mathematical definition of θ (Eqn. 28.1) is therefore equivalent to the temperature scale established as above.

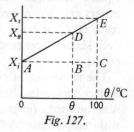

Fig. 127.

A fundamental scale of temperature. Every thermometer temperature scale is arbitrary, since it depends on the properties of a particular material. A fundamental scale would be independent of the properties of particular substances. Such a scale exists—the thermodynamic, or work, scale, based on the theoretical efficiency of a perfectly reversible heat engine—but it is a theoretical scale and cannot be realized in practice. However, this fundamental scale can be shown to be identical with the ideal gas scale. Since an ideal gas is any real gas under conditions of infinitely low pressure (§ 26), the fundamental scale can be arrived at by constructing a gas thermometer, containing a particular gas at a particular pressure, and then making corrections to its indicated scale which are calculated from the known deviations of that gas from the gas laws.

An ideal gas at constant volume would exert zero pressure at the absolute zero of temperature. On the ideal gas scale the absolute zero occurs at $-273.15°C$. The absolute zero may be regarded as a further 'fixed point'. The thermodynamic scale uses as two of its fixed points the absolute zero, and the triple point of water (very nearly the ice point). The thermodynamic scale has its zero at the absolute zero and its unit temperature interval the same as that of the Celsius scale.

The KELVIN is the unit of thermodynamic temperature. It is the fraction $1/273.16$ of the thermodynamic temperature of the triple point of water.

Expressed in the form of Eqn. (28.1), a temperature T in kelvin may be defined by

$$\frac{T}{273.16} = \frac{P_T - P_0}{P_{tr} - P_0} \quad\dots\dots\dots\dots\dots(28.2)$$

where P_T, P_{tr}, P_0 are the pressures of an ideal gas at constant volume, at the given temperature T, the triple point of water, and the absolute zero, respectively.

TYPES OF THERMOMETER

Mercury-in-glass thermometer. A temperature θ on the mercury-in-glass Celsius scale is defined by Eqn. (28.1) where the property X is the length of the mercury column in the particular glass.

Errors are (1) Lack of uniformity of bore of tube;
(2) Changes of zero—glass contracts slowly over years;
(3) Error of exposed mercury thread, which is not at the same temperature as the immersed bulb;
(4) Changes of pressure on the bulb causing dilation or contraction.

Constant volume gas thermometer. As explained above, the constant volume gas thermometer can be used to define thermodynamic temperature T in kelvin from Eqn. (28.2).

The bulb A is immersed in a bath at the unknown temperature (Fig. 128). Reservoir B is adjusted to maintain the mercury level C at the constant volume marker D. D is the zero of the fixed scale E at the other end of which runs a vernier scale attached to the barometer tube F, which can be moved vertically. F is adjusted so that its pointer G coincides with the mercury level J. The vernier now reads the difference in levels h in mm between C and J. This is the gas pressure in the bulb A in mmHg.

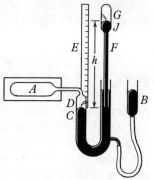

Fig. 128. Constant volume gas thermometer.

The advantages of the gas thermometer are (1) a large range, from the liquefying point of the gas up to about 1500°C; (2) high sensitivity, because of the large pressure coefficient of gases; (3) its reproducibility, i.e. different thermometers will read the same.

Small corrections are needed for the expansion of the bulb, and for the exposed column. The serious disadvantage of the gas thermometer is its large bulb, making it unsuitable for direct measurements of temperature. Its important use is as a standardizing instrument for calibrating other types of thermometer.

Platinum resistance thermometer. This measures temperature by the changes of resistance of a spiral of fine platinum wire, wound on a mica former in a hard glass or metal tube.

Four electrical leads run down the tube. Two of the leads connect to the platinum spiral; the other two, the compensating leads, are joined together (Fig. 129). The two pairs, running side by side, connect into opposite arms of the bridge circuit. Since temperature changes affect the resistance of both pairs equally, their effects cancel out.

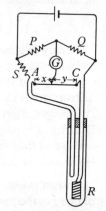

Fig. 129. Platinum resistance thermometer.

The bridge circuit is used to measure the changes in the resistance R of the platinum. S, a known resistance approximately equal to R, is placed as shown. P, Q are equal resistances. If L is the resistance of each pair of leads and x, y are the balancing lengths along the uniform resistance wire AC, of resistance *per unit length r*, then by Eqn. (35.5),

$$\frac{P}{Q} = \frac{S + L + rx}{R + L + ry}$$

giving $R = S + r(x - y)$.

The instrument is calibrated at three fixed points, so that temperatures on the resistance scale can be converted to the gas scale. The success of the instrument as a practical thermometer is due to the reliability of the resistive properties of pure platinum, the ability of the bridge circuit to measure small changes of resistance accurately, and the solution of the problem of the leads resistance.

Thermoelectric thermometer. The hot junction acts as the temperature probe, while the cold junction is maintained at a steady temperature, either in melting ice or at room temperature. For accurate work the thermoelectric e.m.f. is measured with a potentiometer (p. 164). Industrially, a simpler arrangement is used, with a sensitive galvanometer placed directly in the circuit (Fig. 130). This measures the *current*—which is proportional to the e.m.f. generated provided that the total circuit resistance remains constant. A 'ballast'

29 . DEVIATIONS FROM THE GAS LAWS

resistance is incorporated, which is large compared with any variations in leads resistance which may occur due to temperature changes. Each outfit is calibrated individually against a standard thermometer, and the galvanometer scale is marked directly in temperature units.

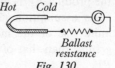

Ballast
resistance

Fig. 130.

Comparison of resistance thermometer and thermocouple. The former is extremely accurate over a wide range of temperatures. Its main drawback is its low thermal conductivity and high thermal capacity which make it unsuitable for measuring changing temperatures. It requires some skill in use and is primarily a laboratory instrument.

The thermocouple has an equivalent wide range but not quite the same accuracy. It is ideally suited to measure varying and localized temperatures. Used extensively in industry owing to its ease of operation, robustness, and ability to give a direct reading.

29. Deviations from the Gas Laws

Deviations by permanent gases. Amagat and others investigated the deviations from Boyle's law by the so-called 'permanent' gases (air, O_2, N_2, H_2, He, etc.) at pressures of up to 3000 atmosphere. It was found that for each gas there is a BOYLE TEMPERATURE —a temperature at which Boyle's law is closely obeyed up to high values of pressures. Above and below this temperature the curves are as shown (Fig. 131).

The curves can be explained in terms of the assumptions made for an ideal gas (§ 26):—

As assumption (2) breaks down with increasing pressure, the increasing attraction between molecules causes the ob-

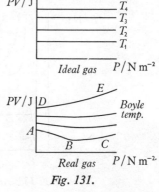

Fig. 131.

served pressure P to be *less* than that of the ideal gas. Consequently the product PV is decreased by this effect.

As assumption (3) breaks down with increasing pressure, the finite size of the molecules causes the observed volume V to be *greater* than that of the ideal gas. This effect increases the product PV.

The two effects thus act in opposition as the pressure is increased. Between A, B (Fig. 131) the former effect exceeds the latter; between B, C the reverse is the case. Above the Boyle temperature, in the curve DE, the latter effect is always the greater, so that PV is always increasing.

Andrews' experiments with CO_2. The object of the experiment was to investigate the P, V, T behaviour of CO_2 in the region of its critical temperature and at high pressures. P-V curves were obtained at various temperatures and plotted as isothermals (Fig. 132*a*).

The volume V of the CO_2 was measured by the length x of the CO_2 column (Fig. 132*b*). Its pressure P measured by the length y of the air column—the volume and pressure of air are related in a known way, and this column was previously calibrated using known pressures. The pressure of the system was varied by screw plungers, the water communicating the hydrostatic pressure to all parts equally.

Results obtained. The 50°C isothermal approximates to a Boyle's law hyperbola, but at lower temperatures a pronounced kink develops. Considering the 20°C isothermal in detail:

In the region AB the vapour is unsaturated, and obeys Boyle's law approximately. At B the vapour becomes saturated, at its saturated vapour pressure. A further decrease in volume results in condensation, and in the region BC the vapour exists in equilibrium with its

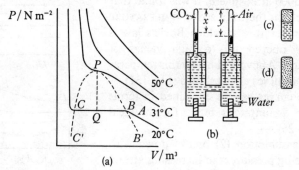

Fig. 132. Andrews' experiment with CO_2.

own liquid. At C the whole space is occupied by liquid. Further decrease in volume results in a rapid increase in pressure, as a liquid is almost incompressible.

Conclusions. Two broken lines $C'C$ (drawn through all points C) and $B'B$ (drawn through all points B) meet at a point P. It is only in the region bounded by $C'CPBB'$ that the CO_2 is heterogeneous, i.e. has a surface separating liquid and vapour states (Fig. 132c). At all points outside this region the CO_2 is homogeneous, i.e. it is either wholly liquid or wholly gaseous (Fig. 132d). Thus it is only at temperatures below that of the critical isothermal (31°C) passing through P that the substance shows at any stage the characteristics of a liquid, which has a surface, as opposed to a gas, which occupies the whole of the containing vessel. The point P is called the 'critical point'.

The experiment thus establishes the idea of a 'critical temperature' (in this case 31°C) above which the substance cannot be liquefied (§ 27).

A further interesting and instructive point emerges. If the temperature of the liquid-vapour at Q is increased at constant volume, the pressure increases, following the broken line QP. Evaporation takes place, the vapour becoming more dense as the temperature is raised. As P is approached, the liquid-vapour boundary becomes indistinct, and at P it disappears. The vapour density is now equal to the liquid density, and there is no distinction between the two states.

Part IV
Sound

30. Wave Motions in General

Definitions of terms

The DISPLACEMENT (x or y) of an oscillating particle is its distance from its mean or rest position at any instant.

The AMPLITUDE (a) is the maximum value of the displacement.

The PERIOD (T) is the time for one complete oscillation or cycle.

The FREQUENCY (f) is the number of complete cycle per unit time.

The ANGULAR FREQUENCY or ANGULAR VELOCITY of a s.h.m. $y = a \cos \omega t$ is the constant ω in radian per second. (See § 4.)

The WAVELENGTH (λ) is the distance between two successive vibrating particles which are in the same phase.

The PHASE of a vibrating particle is the fraction of a period or cycle that has elapsed since the particle passed some fixed point in the cycle.

The PHASE ANGLE between two out-of-phase s.h.m.s, $y = a \cos \omega t$ and $y = a \cos (\omega t + \phi)$, is the angle ϕ in radian.

A WAVEFRONT is a line or surface, all particles on which are vibrating in the same phase.

The WAVE VELOCITY (v) is the velocity of advance of the wavefronts.

The PARTICLE VELOCITY (dx/dt or dy/dt) is the actual velocity of a particle in the wave at any instant.

Classification of waves

A wave motion arises from the to and fro motions of the individual particles of a medium. After the initial transient wave the particles quickly settle down into a definite pattern of vibrations which repeats itself at equal intervals in both space and time. Waves are classified according to the direction of these individual vibrations:—

(A) *Transverse* waves, in which the particles vibrate in a direction perpendicular to the line of advance of the wave.

(B) *Longitudinal* waves, in which the particles vibrate in the same direction as the line of advance of the wave.

The resulting wave may be one of two distinct types:—

(A) *Progressive* waves, in which all the particles have similar vibrations, but with a progressive change of *phase* with distance from the source of the disturbance. A waveform therefore exists, which

136

appears to be propagated steadily forward at a velocity v, the wave velocity (Figs. 133, 4).

(B) *Stationary* waves, in which the particles have similar vibrations, but with a progressive change of *amplitude* with distance along the line of propagation. NODES (points of zero amplitude) alternate with ANTINODES (points of maximum amplitude). A waveform is therefore present, but it does not appear to progress (Fig. 133, 4).*

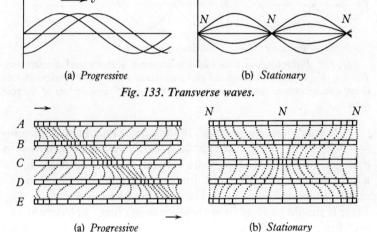

(a) *Progressive*   (b) *Stationary*

Fig. 133. Transverse waves.

(a) *Progressive*   (b) *Stationary*

Fig. 134. Longitudinal waves. A shows the positions of 16 particles in a wave, at a given instant; B shows the positions of the same particles, one quarter of a cycle later; and so on. The five strips thus represent the passage of one complete cycle. The dotted lines trace the motions of the individual particles.

Note how the out-of-phase oscillations in the progressive wave cause the compression to progress from left to right. In the stationary wave, the compressions appear and disappear at the displacement nodes.

Thus there are four categories of waves:—

transverse progressive	e.g.	water waves, light
„ *stationary*	„	vibrations of stretched strings
longitudinal progressive	„	sound
„ *stationary*	„	vibrations of air columns

Longitudinal waves are rather difficult to visualize. Fig. 134 shows one complete cycle of a progressive and a stationary wave of this type.

* Note that the wavelength is the distance between *alternate* nodes (or antinodes).

SOUND

Phase relationships

(1) *Displacement, velocity and acceleration.* When the particle displacement is maximum, its velocity is zero, and its acceleration is maximum in the opposite sense. Displacement and acceleration are therefore in opposite phases, and velocity is in the phase midway between (Fig. 135).

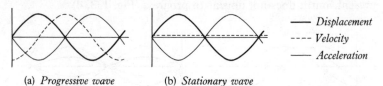

(a) *Progressive wave*   (b) *Stationary wave*

Fig. 135. Instantaneous values of displacement, velocity and acceleration. These apply to both longitudinal and transverse waves. The student should draw similar graphs for one-quarter, one-half and three-quarters of a cycle later.

(2) *Displacement and pressure.* In the case of longitudinal, or compression, waves, the particle displacements give rise to variations in pressure. In Fig. 136a all particles between *A, B* are displaced to the left, between *B, C* to the right. Pressure maxima occur at *A* and *C*, and a minimum occurs at *B*. The opposite phase is shown in Fig. 136b. It should be clear from these diagrams that displacement and pressure are out of phase by one quarter of a cycle. For stationary waves (Fig. 136c) the displacement nodes are the pressure antinodes, and vice versa.

Experimental properties

All wave motions have certain characteristics in common:—

(1) *Reflection.* When waves are reflected, the angle of incidence is equal to the angle of refraction (§ 18).

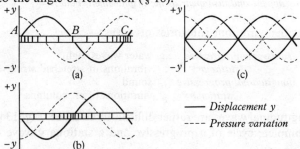

Fig. 136. Instantaneous values of displacement and pressure variation, for longitudinal waves. +y indicates displacements to left, −y to right.

138

(2) *Refraction*. Refraction or bending takes place if the speed of the wave is made to change, and Snell's law is obeyed (§ 18).

(3) *Diffraction*. When waves meet an obstacle, diffraction or spreading out of the wave behind the obstacle takes place, to an extent depending on the size of the obstacle relative to the wavelength (§ 16).

(4) *Interference*. Waves can interfere constructively or destructively (§ 16) although two waves will pass through each other without being affected beyond the region of overlap.

Proof of $v = f\lambda$

A wave progressing to the right (Fig. 137) passes point A and arrives at B one second later. The distance per second AB is therefore v, where v is the wave velocity. But f waves have passed A in one second, where f is the frequency. Each wave has a wavelength λ, so the distance per second AB is also equal to $f\lambda$.

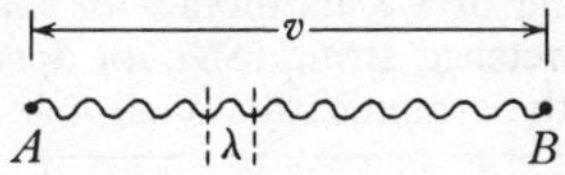

Fig. 137.

$$\therefore \qquad v = f\lambda \qquad \dots\dots\dots\dots\dots\dots\dots\dots\dots(30.1)$$

31. Sound Waves

Sound is produced by a *vibrating* body, e.g. a tuning fork. It is transmitted by longitudinal vibrations of the particles of a material medium—solid, liquid or gas; unlike light, it cannot be transmitted through a vacuum. The human ear can detect these vibrations as sound only if the frequency lies between certain limits—about 30 to 20,000 Hz for the average ear.

We can sense, in a musical note, the characteristics of *pitch*, *tone quality* and *loudness*. Each of these factors is linked with a measurable quantity in the wave:

Pitch. The PITCH of a note is determined by its *frequency*, higher notes having higher frequencies. The intervals of the musical scale are determined by frequency *ratios*. The frequency is doubled at the octave, and the ratios for the notes doh, me, soh, doh, are 4 : 5 : 6 : 8 (see Table, p. 140). The complete scale can be built up from these simple ratios.

139

SOUND

Tone quality. A note is seldom a pure tone, of one frequency only, and usually contains many overtones. The audible overtones in a note may number 20 or 30. The TONE QUALITY or TIMBRE of the note is determined by the relative strengths of these various overtones present with the fundamental.

The FUNDAMENTAL is the component of lowest pitch, and usually of greatest strength, present in a musical note.

A HARMONIC is a note whose frequency is an exact multiple of that of the fundamental. The fundamental itself is the 1st harmonic.

OVERTONES are those harmonics, of higher pitch than the fundamental, which are actually present with it in a given note.

Table III—Harmonics and Overtones

The first 8 harmonics of *middle C*, and the overtones produced by a stretched string (*SS*), an open pipe (*OP*), and a stopped pipe (*SP*). The 7th harmonic is between *A* and *B flat* on the scale.

Notes of musical scale	Frequency Hz	Harmonics	Overtones		
			SS	OP	SP
doh C'''	$8f = 2048$	8th	7th	7th	—
—	$7f = 1792$	7th	6th	6th	3rd
soh G''	$6f = 1536$	6th	5th	5th	—
me E''	$5f = 1280$	5th	4th	4th	2nd
doh C''	$4f = 1024$	4th	3rd	3rd	—
soh G'	$3f = 768$	3rd	2nd	2nd	1st
doh C'	$2f = 512$	2nd	1st	1st	—
doh C	$f = 256$	1st	fundamental		

Intensity, relative intensity and loudness

The INTENSITY (I) of a sound wave crossing an area in space is the rate of flow of energy per unit normal area. *Unit:* $W\ m^{-2}$.

It is a quantity measured by instruments. The intensity is proportional to the square of the amplitude of the vibrations.

Relative intensity. The ear, however, does not detect equal increments of intensity as equal increments of loudness, but becomes less sensitive with increasing intensity, approximately on a logarithmic scale. In this way the ear can deal with an enormous range of intensities. Relative intensity, although still a quantity measured by instruments, has its unit the decibel defined so that increments in decibel correspond roughly to increases in physical sensation:

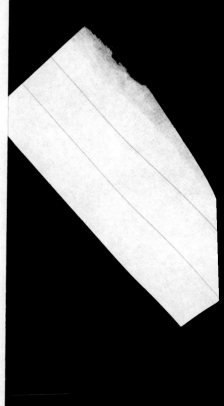

Two sounds of intensities I_1, I_2 are said to differ in intensity by x DECIBEL,* where

$$x = 10 \log_{10} \frac{I_2}{I_1} \quad \ldots\ldots\ldots\ldots\ldots\ldots\ldots\ldots\ldots(31.1)$$

If I_1 is the intensity at the 'threshold of audibility' of a note of the same frequency, x is said to be the intensity of I_2 in 'decibel above threshold'. The threshold of audibility varies with the individual but is taken to occur at an intensity of 10^{-12} W m^{-2} for a sound of frequency 1000 Hz.

Loudness. A further complication is the different sensitivity of the ear at different frequencies. The decibel state takes no account of this, so does not approximate to loudness sensations when different frequencies are involved. Loudness is subjective, and requires for its measurement the judgement of an 'average observer'. To determine loudness, the given sound is compared with a standard note of 1000 Hz of known, variable intensity on the decibel scale for 1000 Hz. The standard note is adjusted until the two are judged to be of equal loudness. Then

the LOUDNESS of the given sound is numerically equal to the intensity in decibel of the standard note of 1000 Hz judged equally loud.

Velocity of transverse progressive waves along a stretched string. Consider a crest of a transverse wave travelling from left to right along a string with a velocity v. Alternatively, we can consider the waveform as remaining stationary and the string itself moving with speed v in the reverse direction, round the curve of the waveform, as in Fig. 138a. At the crest of the wave the string is thus travelling round the arc of a circle of radius r with a velocity v, and consequently has an acceleration v^2/r towards the centre O (Fig. 138b).

The force on an element AB of the string required to maintain this acceleration is provided by the vertical components of the tensions T in the string at A and B. Each component is $T \sin \frac{1}{2}\theta$ or $T \cdot \frac{1}{2}\theta$ (as θ is small). Resultant force towards $O = T\theta$.

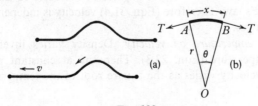

Fig. 138.

* For example, an increase in intensity 1000 times is an increase of 30 decibel.

141

SOUND

If m is the *mass per unit length* of the string, Mass of element $AB = mx = mr\theta$.

Therefore, by Newton's second law $(F = Ma)$, $T\theta = mr\theta v^2/r$,

or
$$v = \sqrt{\frac{T}{m}} \quad\text{...........................}(31.2)$$

Stationary waves in a stretched string are the result of two such progressive disturbances travelling at velocities v in opposite directions.

Velocity of sound waves through a medium. This is determined primarily by the properties of the medium itself—its elastic properties (which promote the movement of the wave) and its inertia (which slows it down). The velocity is not affected by the motion of the source through the medium nor by the frequency nor, appreciably, by the intensity of the wave. We must assume at this stage the general formula.

$$\text{Velocity of sound} = \sqrt{\frac{\text{Appropriate elastic modulus}}{\text{Density of medium}}} \quad (31.3)$$

which for a *gas* becomes

$$v = \sqrt{\frac{\gamma P}{\rho}} \quad\text{...........................}(31.4)$$

where γP is the adiabatic bulk modulus of elasticity, γ the ratio of the principal specific heat capacities, P the pressure of the gas, and ρ its density.

For a *solid*

$$v = \sqrt{\frac{E}{\rho}} \quad\text{...........................}(31.5)$$

where E is Young's modulus, and ρ the density.

Atmospheric effects on sound

Effect of pressure on velocity. As pressure increases so does density, by Boyle's law. Therefore (Eqn. 31.4) velocity is independent of pressure changes.

Effect of temperature on velocity. Density varies inversely as absolute temperature (Eqn. 23.3). Therefore at constant pressure (Eqn. 31.4) velocity varies as the square root of the temperature in kelvin, or

$$\frac{v_2}{v_1} = \sqrt{\frac{T_2}{T_1}} \quad\text{...........................}(31.6)$$

Effect of humidity on velocity. Damp air is less dense than dry air, so velocity increases with increasing humidity.

Effect of wind on audibility. Wind speed increases with height, owing to frictional drag near the surface. Downwind, therefore (Fig. 139*a*), sound veers towards the ground, whereas upwind it tends to be lifted away, making hearing more difficult on the upwind side of the source, at ground level.

Effect of temperature gradient on audibility. On a clear night the earth radiates heat and cools, the air above remaining warm. Sound waves refract in the less dense upper air towards the ground (Fig. 139*b*), and distant sounds are heard distinctly. The opposite conditions, warm surface and cooler upper air, give poor audibility.

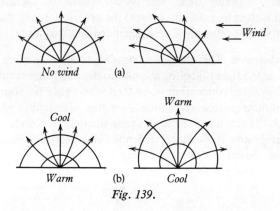

Fig. 139.

Zones of audibility. Since temperature increases with height in the stratosphere, sound should continue to refract until it is actually reflected, returning to the earth at considerable distances from the source, if the initial sound is loud enough. With loud explosions this does happen, and a 'zone of silence' exists between the inner zone receiving sound directly and the outer zone receiving sound by reflection from the stratosphere.

Beats. When two sources of slightly different frequencies f_1, f_2 sound together the interference between the two waves causes distinct throbs, or 'beats', to be heard. It can be shown (§ 33) that

Number of beats per second = Frequency difference, $f_1 - f_2$ ...(31.7)

This phenomenon affords a useful means of tuning two frequencies to unison—by reducing the 'beat frequency' to zero.

143

Forced vibrations and resonance. When a powerful vibration is applied to a body the body begins to vibrate in sympathy. Generally, however, the resulting amplitude is small, because the applied frequency f does not coincide with the natural frequency f_0 of free vibration of the body. Such vibrations are said to be *forced*, and little energy is being absorbed from the applied vibration.

If, however, the applied frequency happens to be equal to the natural frequency of the body, *resonant* vibrations will occur, and the amplitude may build up rapidly to a large value. The body is now able to absorb much of the incident energy.

A graph of absorbed energy against applied frequency gives a resonance peak at $f = f_0$. If the peak is sharp, as with a lightly damped body (e.g. a tuning fork), the system is *selective*. With a heavily damped system (e.g. an air column) the peak is flat and the system is *unselective*: the response is less dependent on frequency.

Reverberation. The REVERBERATION TIME (T) of a room has been defined as the time taken for a sound to die away to one millionth of its intensity (to reduce, that is, by 60 decibel) after the source ceases. The optimum reverberation time is a matter of opinion and of the use to which the room is put, but is approximately 1 second.

Investigations show that T depends almost entirely on A, the total absorbing power of the walls, ceiling and floor, and upon the volume V of the room, where

$$T = \frac{kV}{A} \qquad \dots\dots\dots\dots\dots\dots\dots\dots(31.8)$$

the constant k depending only on the units used.

The quantity A is obtained from $A = a_1s_1 + a_2s_2 + \dots$, where s is the area of a material and a its absorptive power. The ABSORPTIVE POWER of a material is the ratio of its absorption to that of an equal area of open window.

The material of the seats in an auditorium should be such that the absorption is the same whether the seat is being occupied or not. As both V and A are known the reverberation time of a hall can be estimated by the architect before the building is erected.

Vibrations of stretched strings

A stretched string by itself would vibrate for some time but emit little sound. When attached to the body of a stringed instrument, its vibrations are communicated to the sound-board and air in the box, and these are set into *forced* vibration. The larger vibrating area

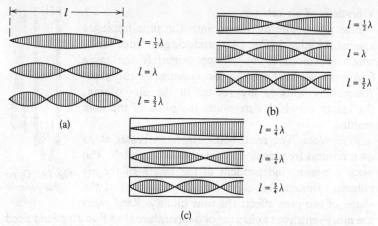

Fig. 140. The first three modes of vibration of (a) a stretched string, (b) an open pipe, and (c) a stopped pipe.

causes appreciable disturbance of the air, and the sound is greatly enhanced. The energy of the string is now being absorbed quickly, and its oscillations soon die away.

When the stretched string is plucked, bowed or struck, two disturbances travel from the point in opposite directions, and are reflected from the ends. The two disturbances travelling in opposite directions combine to form a complex stationary wave, made up of many of the possible modes of vibration, so producing a fundamental with many overtones. The quality of the note produced depends upon the manner of excitation, the position of excitation along the wire (overtones with nodes at this point are thereby excluded), upon any damping at points along the wire (promoting overtones with nodes at these points), and on the nature of the sound-board to which the string is attached.

Displacement nodes must exist at the ends, and the possible modes of vibration are deduced by drawing diagrams (Fig. 140). For the first mode $l = \frac{1}{2}\lambda$, and the fundamental frequency f is therefore given by

$$f = \frac{v}{\lambda} = \frac{1}{2l}\sqrt{\frac{T}{m}} \qquad \qquad (31.9)$$

(Eqns. 30.1 and 31.2). The string can sound all the harmonics f, $2f$, $3f$, etc., and the frequency f_n of the nth harmonic is given by

$$f_n = \frac{n}{2l}\sqrt{\frac{T}{m}} \qquad \qquad (31.10)$$

145

SOUND

Vibrations of air columns

Flue pipes. The air blown into the pipe impinges against the edge E (Fig. 141) producing an 'edge tone', which is voiced to be of approximately the same frequency as that of the air column in the pipe. *Resonant* vibrations are set up in the air column, the length of which determines the pitch of the note emitted.

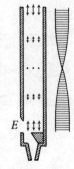

Fig. 141. Open flue pipe.

Reed pipes. The reed, being strong, vibrates at its own natural frequency and determines the pitch. The pitch is almost independent of the length of the air column—though this contributes resonance, and the shape of the pipe affects the tone quality. Reed pipes are more sensitive to changes of temperature than flue pipes and need tuning more often. Tuning is effected by altering the length of the vibrating reed.

Open and stopped pipes. Both ends of an open pipe are open to the air, so presumably pressure nodes, and therefore displacement antinodes, exist at these points. This can be shown experimentally to be so. For a stopped pipe, the closed end must be a displacement node. The possible modes of vibration of the air column imposed by these end-conditions can be deduced by drawing diagrams (Fig. 140). It is seen that an open pipe can sound all the harmonics of the fundamental, a stopped pipe only the odd harmonics. The tone of the latter is therefore less rich; however, the length of the stopped pipe necessary to produce a given pitch is only half that of the open pipe. The fundamental frequency f of an open pipe is given by

$$f = \frac{v}{\lambda} = \frac{v}{2l} \qquad \qquad (31.11)$$

and of a stopped pipe by

$$f = \frac{v}{4l} \qquad \qquad (31.12)$$

where v is the velocity of sound in the pipe. Owing to friction, v is rather less than the velocity of sound in free air.

Doppler effect

The DOPPLER EFFECT is the apparent change in frequency of a wave due to the motion of observer and source.

Let the observer be denoted by O and the source by S. When O and S are stationary in still air (Eqn. 30.1),

$$V = f\lambda,$$

where V is the velocity of the wave, f the frequency of the source, and λ the wavelength.

(1) *S stationary, O approaching S at velocity u.* The velocity of the wave relative to O is increased from V to $(V + u)$. The wavelength λ is unchanged. The observed frequency is increased from f to f', where

$$\frac{f'}{f} = \frac{(V + u)/\lambda}{V/\lambda} = \frac{V + u}{V} \quad \text{.................(31.13)}$$

(2) *O stationary, S approaching O at velocity v.* The velocity V of the wave is unchanged by the motion of the source (p. 142). The wavelength, however, is reduced; this is shown as follows. After one second, a wave from a stationary source S reaches P, where distance per second $PS = V$ (Fig. 142a).

Now if, instead, S is approaching P at velocity v, S moves in this second a distance v to S' (Fig. 142b). Thus $PS' = (V - v)$. In both cases f waves have been emitted, but in the latter case these occupy a smaller distance, and the wavelength is correspondingly reduced, in the ratio

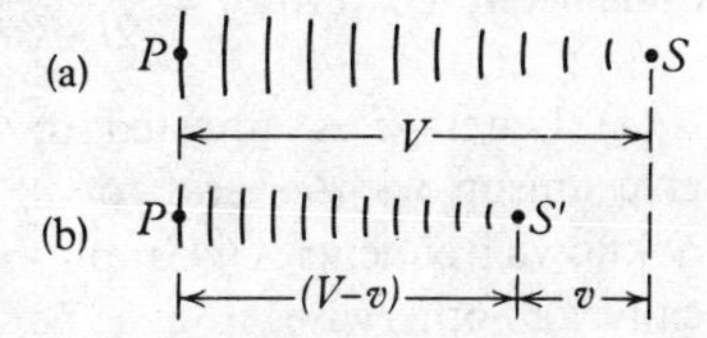

Fig. 142. *Effect of motion of source on wavelength.*

$$\frac{\lambda'}{\lambda} = \frac{V - v}{V} \quad \text{.............................(31.14)}$$

$$\therefore \quad \frac{f'}{f} = \frac{V/\lambda'}{V/\lambda} = \frac{V}{V - v} \quad \text{.....................(31.15)}$$

(3) *O, S approaching each other at velocities u, v, respectively.* The motion of O increases the velocity of the wave past O, and the motion of S decreases the wavelength to λ', as above.

$$\therefore \quad \frac{f'}{f} = \frac{(V + u)/\lambda'}{V/\lambda} = \frac{V + u}{V - v} \quad \text{.................(31.16)}$$

(4) *Effect of wind blowing from S towards O at velocity w.* The wind has the effect of increasing the velocity of sound in this direction from V to $(V + w)$. This affects both the velocity past O and the wavelength.

$$\therefore \quad \frac{f'}{f} = \frac{(V + w) + u}{(V + w) - v} \quad \text{.................(31.17)}$$

This is the general formula for sound waves. When applying the Doppler effect to light waves, use Eqn. (31.14). Note that u, v, w are vector quantities: if the movements are not in the directions specified above, account must be taken of the algebraic signs of these quantities; if the movement is at an angle to the line OS, the component in the appropriate direction must be used.

32. Experiments

Sonometer. To verify $f = \dfrac{1}{2l}\sqrt{\dfrac{T}{m}}$ (Eqn. 31.9). See Fig. 143.

(1) Length l is adjusted to give unison with several forks of known frequencies f. Graphically or otherwise it can be shown that $f \propto 1/l$, with tension constant.

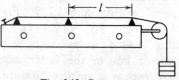

Fig. 143. Sonometer.

(2) Length l is adjusted to give unison with a certain fork for various tensions T. It can be shown that $l \propto \sqrt{T}$, for a given frequency.

Note that in each case a variable changing in finite steps is matched against the only continuous variable, l.

Tuning to unison is achieved either by observing beats between tuning fork and wire, sounded together; or by placing a small paper rider on the wire and adjusting the length until the rider is thrown off by resonance when the vibrating fork is pressed on the sonometer board.

The accuracy of the experiment is limited by friction over the pulley, which reduces the tension. The sonometer can, however, be used in a near-vertical position.

Melde's experiment. Illustrates the laws of vibrating strings.

A frequency is applied to the string: either by attaching one end to the prong of a tuning fork; or by placing the stretched wire at right angles to a magnetic field and passing an alternating current along it.

The tension T, or length l, is adjusted until n resonance loops appear in the string. The applied frequency then equals the natural frequency of vibration of the string in that mode, given by $f = \frac{n}{2l}\sqrt{\frac{T}{m}}$(Eqn. 31.10). It can be demonstrated that $1/n \propto \sqrt{T}$, with the length constant.

Fig. 144 shows the prongs vibrating in a direction perpendicular to the string; in this case $f = F$, where F is the frequency of the fork. If the prongs are arranged to vibrate in line with the string, the applied frequency $f = \frac{1}{2}F$ (this can be shown by diagrams). With a.c. excitation, the applied frequency is equal to the frequency F of the mains.

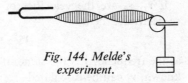

Fig. 144. Melde's experiment.

Resonance tube. To find velocity of sound v and end correction c.

The length l is adjusted by raising the tube until the sound swells out when a vibrating tuning fork of known frequency f is held over the end of the tube (Fig. 145a). The mean of several readings gives l_1.

By raising the tube further, the second position of resonance is then found with the same fork (Fig. 145b). Let this new length found be l_2.

At each resonance position the air column vibrates at the frequency of the fork. A displacement node occurs at the closed end and an antinode at a small distance c above the open end, where c is called the *end correction* and is approx. 0.6 of the radius of the tube.

Fig. 145 shows that $l_2 - l_1 = \lambda/2$, thus eliminating the end correction.

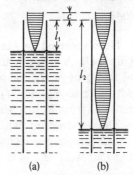

Whence $v = f\lambda = 2f(l_2 - l_1)$.

(a)   (b)

Fig. 145. Resonance tube.

Also, we can calculate c from $l_1 + c = \lambda/4$.

The experiment is repeated with other forks of known frequencies.

Kundt's dust tube. A rosined cloth is drawn along the rod which is firmly clamped at its centre (Fig. 146). A shrill sound indicates that the rod is set in longitudinal vibration, with a node at the centre and antinodes

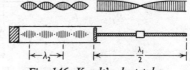

Fig. 146. Kundt's dust tube.

149

at the ends. If λ_1 is the wave-length of the sound waves in the rod, it is seen that λ_1 is twice the length of the rod.

The length of glass tubing between piston and closed end is adjusted until striations in a periodic pattern suddenly appear in a thin layer of cork dust spread evenly along the tube. The piston, vibrating longitudinally at a frequency f, has produced resonant stationary waves of the same frequency f in the air inside the tube. The periodic pattern in the tube repeats itself every *half* wavelength of the air waves, so their wavelength λ_2 can be measured.

If v_1, v_2 are the velocities of sound waves in the rod, and air, respectively, then

$$v_1\lambda_2 = v_2\lambda_1,$$

as the frequency f is the same. If v_2 at the prevailing temperature is known, v_1 can be calculated.

As $v_1 = \sqrt{\dfrac{E}{\rho}}$ (Eqn. 31.5), Young's modulus E for the material of the rod can be calculated, if a separate specimen is available for measuring the density ρ.

If the air in the tube is replaced by CO_2 and the experiment repeated, the velocity of sound in the gas can be found from $v_2\lambda_3 = v_3\lambda_2$.

Hebb's method for v in free air. R_1, R_2 are large parabolic reflectors (Fig. 147). A high-pitched source S of known frequency f is sounded at the focus of R_1, the sound being picked up by two microphones, M_1 close to S, and M_2 at the focus of R_2. The sound from S is reflected from R_1 in a parallel beam, falls on R_2, and is focused to M_2.

The responses in M_1, M_2 are led to identical primary windings of a transformer, the secondary being connected to an earpiece E. R_2, together with M_2, is moved along the line R_1R_2 until a maximum

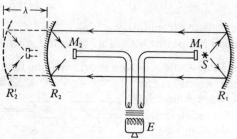

Fig. 147. Hebb's method for v.

sound is heard in E, at which point the waves striking M_1 and M_2 interfere constructively. As R_2 is moved, successive maxima and minima are heard, the distance between any two adjacent maxima (or minima) corresponding to a path difference $R_2R_2' = \lambda$ (Eqn. 16.1).

Knowing λ and f, the velocity v of sound in air at the prevailing temperature is found from $v = f\lambda$.

Measurement of frequency f by siren. To find the frequency of a given musical note a source of variable, known frequency, such as a siren, can be used.

The siren is adjusted to unison with the note by the use of beats. If this cannot be done exactly, and a number of beats per second b remains when the siren frequency is F, then the unknown frequency $f = F \pm b$ (Eqn. 31.7). The ambiguity is removed by increasing F very slightly. If b is thereby decreased, F is approaching f and is therefore less than f, so $f = F + b$. If b increases, $f = F - b$.

The siren produces puffs of air as holes in a revolving wheel pass in front of a blast of air. The frequency F is therefore nN, where n is the revolution per second and N the number of holes in the wheel. Using a stopwatch and revolution counter, n can be found.

f of tuning fork by falling plate method. A smoke-blackened glass slide is allowed to fall freely under gravity in front of the vibrating tuning fork, to one prong of which is attached a bristle which just touches the falling plate. The wavy trace made on the plate (see Fig. 33a, Fletcher's trolley) is examined under a travelling microscope and the lengths s_1, s_2 of two successive sets of n complete waves each are measured.

Using the constant acceleration formulae, exactly as in Fletcher's trolley experiment ($\S 8$), we obtain $s_2 - s_1 = gt^2$. Hence t, the time taken for n vibration. The frequency f of the fork is given by $f = n/t$.

33. Mathematical Treatment

An analysis of some of the foregoing phenomena is included below for the more mathematical student. The treatment is for simple harmonic waves.

Equation of a plane progressive wave. The equation

$$y = a \cos \omega t \quad \text{......................(33.1)}$$

represents the simplest form of vibration—s.h.m.—at a point. The equation

$$y = a \cos(\omega t - \phi) \quad \text{....................(33.2)}$$

represents the same vibration with the introduction of a *phase angle* ϕ.

In a wave motion, this phase angle is a function of the distance x along the wave—in such a way that when $x = 0$, λ, 2λ, etc. then $\phi = 0$, 2π, 4π, etc. In other words

$$\frac{\phi}{x} = \frac{2\pi}{\lambda} \quad \text{..........................(33.3)}$$

Substituting for ϕ we have the equation of a plane progressive wave. This equation* gives the displacement y at all times t for all distances x along the wave:

$$y = a \cos\left(\frac{2\pi}{\lambda}x - \omega t\right) \quad \text{....................(33.4)}$$

Period $T = \dfrac{2\pi}{\omega}$; frequency $f = \dfrac{1}{T} = \dfrac{\omega}{2\pi}$, where ω is the angular frequency; wave velocity $v = f\lambda$. Substituting these, we can express the wave equation in various forms:

$$y = a \cos 2\pi\left(\frac{x}{\lambda} - \frac{t}{T}\right) = a \cos \frac{2\pi}{\lambda}\left(x - vt\right), \text{ etc. ...(33.5)}$$

Constructive and destructive interference. The two equations $y_1 = a \cos \omega t$ and $y_2 = a \cos(\omega t - \phi)$ represent two waves out of phase by an angle ϕ but otherwise identical. By the principle of superposition (p. 76) the resulting wave is

* The equation is usually written as shown. Note that $\cos(A-B) = \cos(B-A)$.

152

$$y = y_1 + y_2$$
$$= a[\cos \omega t + \cos(\omega t - \phi)]$$
$$= 2a \cos \tfrac{1}{2}(\omega t - \omega t + \phi) \cos \tfrac{1}{2}(\omega t + \omega t - \phi)*$$
$$= 2a \cos \tfrac{1}{2}\phi \cdot \cos(\omega t - \tfrac{1}{2}\phi).$$

(1) When the two components are in phase, $\phi = 0$, 2π, etc. and the resultant is $y = 2a \cos \omega t$. This is constructive interference, the amplitude being $2a$.

(2) When exactly out of phase, $\phi = \pi$, 3π, etc. and the amplitude $2a \cos \tfrac{1}{2}\phi$ is zero at every instant. This is destructive interference.

(3) When the phase difference is one quarter of a cycle the resultant is $y = \sqrt{2}a \cos(\omega t - \pi/4)$. This has an amplitude of $\sqrt{2}a$ and a phase midway between those of the components.

The resulting waves in the three cases are shown in Fig. 88, § 16.

Beats. The two equations $y_1 = a \cos 2\pi f_1 t$ and $y_2 = a \cos 2\pi f_2 t$ represent two waves of slightly different frequencies f_1 and f_2. The resulting wave is

$$y = y_1 + y_2$$
$$= a[\cos 2\pi f_1 t + \cos 2\pi f_2 t]$$
$$= 2a \cos 2\pi \frac{f_1 - f_2}{2} t \cdot \cos 2\pi \frac{f_1 + f_2}{2} t*$$

which is a modulated wave (Fig. 148) the beat frequency being determined by the function $\cos 2\pi \frac{f_1 - f_2}{2} t$ which has a frequency of $\frac{f_1 - f_2}{2}$. However, as *two* beat maxima occur in each cycle the beat frequency is twice that of this function.

$$\therefore \qquad \text{Beat frequency} = f_1 - f_2.$$

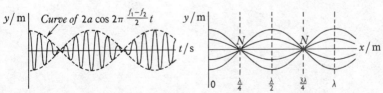

Fig. 148. Formation of beats.   Fig. 149. Stationary waves.

* Using $\cos A + \cos B = 2 \cos \tfrac{1}{2}(A - B) \cos \tfrac{1}{2}(A + B)$.

153

SOUND

Stationary waves. The two equations $y_1 = a \cos \frac{2\pi}{\lambda}(x + vt)$ and $y_2 = a \cos \frac{2\pi}{\lambda}(x - vt)$ represent two identical progressive waves travelling in opposite directions. Their resultant is

$$y = y_1 + y_2 = 2a \cos \frac{2\pi}{\lambda}x \,.\, \cos \frac{2\pi}{\lambda}vt *$$

which is a stationary wave (Fig. 149). All particles are oscillating in the same phase, and the amplitude is zero at $x = \frac{1}{4}\lambda, \frac{3}{4}\lambda$, etc., where the nodes occur.

* Using $\cos A + \cos B = 2 \cos \frac{1}{2}(A - B) \cos \frac{1}{2}(A + B)$.

Part V
Electricity

34. Current Electricity Theory

Definitions and laws

OHM'S LAW. The current I through a metallic conductor is proportional to the potential difference V across its ends, provided that the temperature remains constant.

Thus V/I is a constant. This constant is called the *resistance* (R) of the metallic conductor,

or $$\frac{V}{I} = R \quad \dots\dots\dots\dots\dots\dots\dots\dots\dots(34.1)$$

THE OHM. A metallic conductor has a resistance of 1 ohm if a current of 1 ampere flows when a potential difference of 1 volt is maintained across its ends.

The RESISTIVITY (ρ) of a material is defined by the equation

$$R = \frac{\rho l}{A} \quad \dots\dots\dots\dots\dots\dots\dots\dots\dots(34.2)$$

where R is the resistance of a uniform specimen of length l and cross-sectional area A. *Unit:* ohm metre.

The resistivity is thus numerically the resistance of a specimen of unit length and unit cross-sectional area.

The TEMPERATURE COEFFICIENT OF RESISTANCE (a) of a material is its fractional increase in resistance at 0°C, per degree rise in temperature. *Unit:* per kelvin.

Over a fairly small temperature range a may be considered constant, so that

$$R = R_0(1 + a\theta) \dots\dots\dots\dots\dots\dots\dots(34.3)$$

where R_0, R are the resistances at 0°C and temp. θ in °C.

THE AMPERE. (Defined on p. 177.)

The COULOMB is that quantity of electricity which passes when a current of 1 ampere flows for 1 second.

Thus $\qquad$ COULOMB = AMPERE × SECOND

or $$Q = It \quad \dots\dots\dots\dots\dots\dots\dots\dots(34.4)$$

THE VOLT. A potential difference of 1 volt exists between two points if 1 joule of work is done between the points when 1 coulomb of electricity passes.

156

34 . CURRENT ELECTRICITY THEORY

The ELECTROMOTIVE FORCE (E.M.F.) of a source of electrical energy is the energy it supplies per unit quantity of electricity passing through it. *Unit:* joule per coulomb, or volt.

The BACK E.M.F. between two points is the electrical energy converted to another form between the points per unit quantity of electricity passing. *Unit:* joule per coulomb, or volt.

The JOULE is the SI unit of *energy*. (Definition, p. 4.)

The WATT is the SI unit of *power*. It is a rate of working of 1 joule per second.

The KILOWATT-HOUR (kWh) is a non-SI unit of *energy*. It is the energy consumed in 1 hour by an electrical appliance working at a power of 1 kilowatt (1000 watt). $1 \text{ kWh} = 3.6 \times 10^6$ joule.

Formulae from definition of volt. By definition of the volt, when a quantity of electricity Q falls through a p.d. V,

$$\text{Work done in joule} = VQ \quad\quad\quad\quad (34.5)$$

If V exists across a resistance R, the energy is dissipated as heat. When a current I flows for a time t a quantity of electricity It passes, and

$$\text{Heat dissipated in joule} = VIt = I^2Rt = \frac{V^2t}{R} \quad\quad (34.6)$$

The *power* equation is

$$\text{Heat per second in watt} = VI = I^2R = \frac{V^2}{R} \quad\quad (34.7)$$

Similarly, by definition of e.m.f., when a current I flows for a time t through a source of e.m.f. E,

$$\text{Energy supplied in joule} = EIt \quad\quad\quad\quad (34.8)$$

Expressing these formulae in another way, we have, by definition of the volt,

$$\text{JOULE} = \text{VOLT} \times \text{COULOMB}$$
$$= \text{VOLT} \times \text{AMPERE} \times \text{SECOND},$$

and, by definition of the watt,

$$\text{JOULE} = \text{WATT} \times \text{SECOND}.$$
$$\therefore \quad\quad \text{WATT} = \text{VOLT} \times \text{AMPERE},$$

or

$$P = VI \quad\quad\quad\quad\quad\quad\quad\quad (34.9)$$

157

ELECTRICITY

The water analogy. The experimental facts are such that a useful analogy can be made between the flow of electricity in a wire and the flow of water in a pipe. The purpose of this analogy is to obtain a clearer picture of the meanings of the terms 'current', 'potential difference' and 'electromotive force'.

Current. The current in the pipe is the rate of flow of water, and could be measured by the number of gallon per second passing any point in the pipe. *The current is the same at all points along the pipe.* Otherwise, if, say, the current entering at A were greater than the current leaving at B, there would be a progressive accumulation of water between A and B.

Similarly, an electric current is a rate of flow of electricity along the wire, measured in coulomb per second, or *ampere*, passing any given point. Ammeters placed at various points in a series circuit all indicate the same current. In a branched circuit, the sum of the currents in the branches is equal to the current in the main circuit, or $I = I_1 + I_2$ (Fig. 151b). Both of these results are to be expected if, indeed, something *is* flowing along the wire.

Potential difference. Water flows in a pipe because different points in the pipe are at different levels. It is not the actual heights of the points above some arbitrary zero (e.g. mean sea level) that matter, but the *difference* in heights, and this determines both the direction and the rate of flow.

Similarly, electricity flows in a wire because different points in the wire are at different 'electric levels' or potentials. It is the difference in potentials, or 'potential difference', between two points which determines both the direction and the rate of flow of electricity in the wire. This analogy—between potential difference and vertical

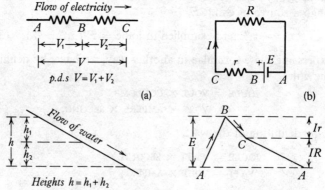

Fig. 150. The water analogy.

158

height—clarifies many points in both current and static electricity.

Since a p.d. is measured *between two points* the voltmeter is connected in parallel across them, whereas a current is measured *at a point* and the ammeter is connected in series there (Fig. 153*a*). In a series circuit, the sum of the p.d.s between *AB* and *BC* is equal to the p.d. between *AC*, or $V = V_1 + V_2$ (Fig. 150*a*).

Electromotive force. It should be realized that every point in a circuit is at some definite 'electric level' or potential. The electricity flows 'down hill' through the external part of the circuit from points of higher potential to points of lower. 'Potential energy' is thus lost, the energy being converted to heat, or some other form. Hence the definition of the *volt* (p. 156).

The function of the battery or other source of e.m.f. in the circuit is thus clearly seen. It is simply to *raise* the potential of the electricity, as it passes through the battery, to its former level, restoring the potential energy lost in the external part of the circuit.

Fig. 150*b* shows a complete circuit and an 'elevation' diagram, with corresponding letters, indicating the electric levels of the various points in the circuit. The internal resistance *r* of the battery is shown separately from the battery itself, and it is seen that part of the potential drop occurs across this internal resistance, so that the p.d. *V* available across the terminals is less than the total e.m.f. *E* provided by the cell. But if no current is flowing (open circuit) the internal potential drop *Ir* is zero, and the p.d. measured across the terminals is the full value of the e.m.f.

A *forward* e.m.f., or 'step up' in potential, thus exists at points in a circuit where electrical energy is being *supplied* to the circuit. A *back* e.m.f., or 'step down' in potential, exists in those parts of the circuit where electrical energy is being *used up*, i.e. converted back to heat, mechanical, or chemical, energy (see, for example, electric motor, p. 195, and electrolysis, p. 199).

Resistances in series. A current *I* flows through two resistances R_1, R_2 in series (Fig. 151*a*). P.d.s V_1, V_2 exist across them, the total p.d. being *V*, where

$$V = V_1 + V_2.$$

But $V = IR$, $V_1 = IR_1$, $V_2 = IR_2$,

applying Ohm's law to each resistance in turn, and to the whole, considered as a single resistance *R*.

Substituting and cancelling *I*,

$$R = R_1 + R_2 \qquad \dots \dots \dots \dots (34.10)$$

159

ELECTRICITY

Resistances in parallel. A p.d. V is applied across two resistances R_1, R_2 in parallel (Fig. 151b). Currents I_1, I_2 flow through them, the total current being I, where

$$I = I_1 + I_2.$$

But $\qquad I = \dfrac{V}{R}, \qquad\qquad I_1 = \dfrac{V}{R_1}, \qquad\qquad I_2 = \dfrac{V}{R_2},$

applying Ohm's law to each resistance in turn, and to the whole, considered as a single resistance R.

Substituting and cancelling V,

$$\frac{1}{R} = \frac{1}{R_1} + \frac{1}{R_2} \quad\dots\dots\dots\dots\dots\dots(34.11)$$

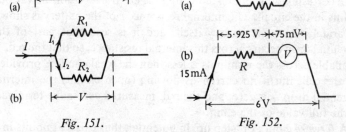

Fig. 151.　　　　　　　　　Fig. 152.

Conversion of ammeters and voltmeters. One type of moving-coil meter commonly used in laboratories is marked '5Ω, 15 mA, 75 mV.' This means that its coil has a resistance of 5 ohm and that it will always register a 'full scale deflection' when a current of 0.015 ampere is passed through it and when a p.d. of 0.075 volt exists across its terminals.

(1) *To convert this instrument into an ammeter reading 0–3 ampere.* A small resistance in parallel is required (Fig. 152a). At full scale deflection only 15 mA must pass through the instrument; the remainder of the 3 A (2.985 A) must pass through the shunt resistance R. Also, a p.d. of 75 mV then exists across its terminals, and the same p.d. exists across the shunt. Applying Ohm's law to the shunt,

$$R = \frac{0.075}{2.985} = 0.0251 \text{ ohm.}$$

(2) *To convert this instrument into a voltmeter reading 0–6 volt.* A large resistance in series is required (Fig. 152*b*). At full scale deflection only 75 mV must exist across the instrument; the remainder of the 6 V (5.925 V) must exist across the series resistance *R*. Also, a current of 15 mA then passes through the instrument, and the same current passes through the bobbin resistance. Applying Ohm's law to the bobbin,

$$R = \frac{5.925}{0.015} = 395 \text{ ohm.}$$

35. Electrical Measurements

Verification of Ohm's law. A moving-coil voltmeter is actuated by a small current passing through it, and is calibrated to read in volt on the assumption that Ohm's law is true. Such an instrument cannot therefore be used, in conjunction with an ammeter, to verify Ohm's law. For this purpose some form of electrostatic voltmeter must be employed, which measures p.d. by the force exerted between static charges on two plates separated by an insulated space, and does not pass a current. Such instruments—quadrant electrometer, attracted disc electrometer, gold leaf electroscope—are inconvenient for ordinary use as the forces involved are very small.

Measurement of resistance by voltmeter and ammeter. The voltmeter reading *V* is divided by the ammeter reading *I* to obtain the resistance *R* (Eqn. 34.1). However, if the instruments are connected as in Fig. 153*a* it is seen that *V*/*I* gives, not the true resistance of *R*, but that of *R* and the voltmeter in parallel. Similarly, with the connections as in Fig. 153*b* *V*/*I* gives, again, not the true resistance *R*, but the resistance of *R* and the ammeter in series.

In either case, therefore, an error is made (which can be corrected by calculation if necessary). Which circuit produces the smaller error will depend upon the relative resistances of voltmeter, ammeter and *R*. In the former arrangement the ideal voltmeter would have infinite resistance and take no current from the circuit. For accurate work a potentiometer is therefore used, since this takes no current (p. 164).

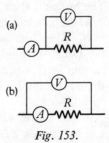

Fig. 153.

ELECTRICITY

Finding *E* and *r* of a cell. The e.m.f. *E* is found directly by measuring the p.d. across the terminals of the cell on open circuit (switch *S* open in Fig. 154). With an open circuit no current flows, so there is no potential drop in the cell itself, and the p.d. across its terminals is the full value of *E*.

The internal resistance *r* is found by taking a series of readings of the p.d. *V* for various values of the current *I* (with switch *S* now closed) and calculating *r* from Eqn. (35.3). Alternatively, readings of *V* can be taken against the resistance *R* and *r* calculated from Eqn. (35.4). The p.d.s can be measured either by voltmeter or by potentiometer (p. 164).

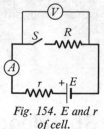

Fig. 154. E and r of cell.

Applying Ohm's law to the complete circuit,

$$E = I(R + r) \dotfill (35.1)$$

and to the external part,

$$V = IR \dotfill (35.2)$$

Subtracting one equation from the other,

$$E - V = Ir \dotfill (35.3)$$

or dividing,

$$\frac{E}{V} = \frac{R + r}{R} \dotfill (35.4)$$

Wheatstone bridge

Proof of formula. P, Q, R, S are the resistances in ohm of the four arms of the bridge (Fig. 155a). These are adjusted in value until the centre-zero galvanometer G shows no deflection. The bridge is then 'balanced'.

At the balance point no current flows in *BC*. Therefore the points *B*, *C* are at the same potential; furthermore, the current in *Q* is equal to the current in *P*, and the current in *S* is equal to the current in *R*.

∴   P.d. across *AB* = P.d. across *AC* and P.d. across *BD* = P.d. across *CD*.

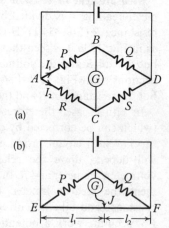

Fig. 155. Wheatstone bridge.

162

Applying Ohm's law to each resistance, where the currents are I_1, I_2 as shown,

$$I_1 P = I_2 R$$

and $$I_1 Q = I_2 S.$$

Dividing one equation by the other, and cancelling I_1 and I_2,

$$\frac{P}{Q} = \frac{R}{S} \quad \dots\dots\dots\dots\dots\dots\dots\dots(35.5)$$

In the *metre bridge* (Fig. 155b) the resistances R, S are in the form of a uniform resistance wire EF of length 1 metre, along which a sliding contact J may be moved. As the wire is uniform the resistance is proportional to length, or $\dfrac{R}{S} = \dfrac{l_1}{l_2}$, where l_1 and l_2 are the lengths, as shown, at the balance point. Substituting for R/S in Eqn. (35.5),

$$\frac{P}{Q} = \frac{l_1}{l_2} \quad \dots\dots\dots\dots\dots\dots\dots\dots(35.6)$$

Post Office Box. This is a convenient form of the Wheatstone bridge, the box itself incorporating three of the resistance arms, and the two switches. In the particular model illustrated (Fig. 156a) the unknown resistance, the galvanometer, and the battery, are connected up as shown. The student should satisfy himself that the circuit is identical with that shown in Fig. 156b, which has the same lettering.

To measure a resistance to two places of decimals:—

(1) Ensure that all plugs are firmly in position. Remove the 10Ω plug from each of the ratio arms AB, BC. Obtain as near a balance as possible by removing plugs from CD. A galvanometer deflection is obtained by holding down the battery key K_1 and tapping lightly the galvanometer key K_2. Suppose the resistance of CD is 43Ω at the balance point. The unknown resistance X is thus 43Ω, to the nearest ohm.

(2) Now remove the 10Ω plug from AB and the 1000Ω plug from

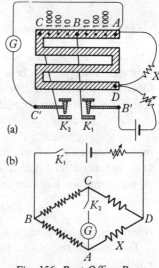

Fig. 156. Post Office Box.

ELECTRICITY

BC. Set 4300Ω in *CD*, and balance again, using a very sensitive galvanometer. Suppose the resistance in *CD* at the balance point is now 4325Ω. Then the unknown resistance *X* is 43.25Ω, now correct to two decimal places.

If the galvanometer is sufficiently sensitive, and a deflection still remains, this may be used to estimate the third decimal place.

THE POTENTIOMETER

Principle. The potentiometer is a form of voltmeter—it measures the p.d. between two points. The probes *A'*, *B'* are connected to the two points in question (contained in a circuit not shown in Fig. 157).

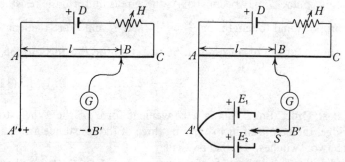

Fig. 157. Principle of potentiometer.   *Fig. 158. Comparing e.m.f.s.*

The cell *D* maintains a constant p.d. across the uniform resistance wire *AC*. This p.d. can be adjusted, within limits, by the rheostat *H*. Since the wire *AC* is uniform the fall of potential along it is uniform. If this is *v* in volt per cm and if the length *AB* is *l* in cm then the p.d. *A* to *B = vl* in volt.

The sliding contact *B* is adjusted until the centre-zero galvanometer *G* shows no deflection, in which case the points *B*, *B'* are at the same potential. The p.d. *A* to *B* then equals the unknown p.d. *A'* to *B'* being applied from some external circuit.

$$\therefore \qquad \text{Unknown p.d.} = vl \dots\dots\dots\dots\dots\dots(35.7)$$

where *l* is the balancing length.

Two conditions must be satisfied for a balance point to be possible:—

(1) the total p.d. available across *AC* must be greater than the p.d. to be measured, and

(2) both p.d.s must be acting in the same direction; this is achieved by ensuring that both *A* and *A'* are connected to the positive terminal.

164

(1) Comparing two e.m.f.s. The two cells of e.m.f.s E_1, E_2 are connected in parallel across $A'B'$ and selected in turn by the two-way switch S (Fig. 158). If the balancing lengths are respectively l_1 and l_2,

$$E_1 = vl_1$$
and
$$E_2 = vl_2.$$

Dividing,
$$\frac{E_1}{E_2} = \frac{l_1}{l_2} \quad\quad\quad\quad\quad\ldots\ldots\ldots\ldots\ldots\ldots\ldots(35.8)$$

Several pairs of readings can be obtained, by adjusting the rheostat H after each pair.

(2) Comparing two resistances. The two resistances R_1, R_2 are connected in *series* with each other, a cell D', and a rheostat H_2. This forms a circuit through which a current I flows (Fig. 159). A wire from each of the points X, Y, Z terminates at a sliding contact which can be pressed on to the potentiometer wire AC. The centre wire from Y is connected through a galvanometer G.

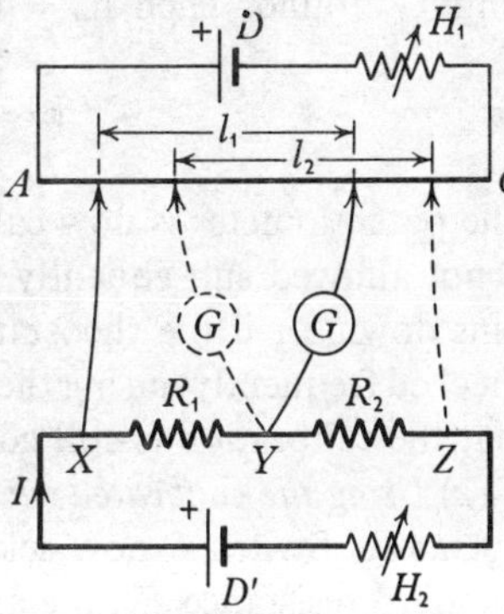

Contact is first made on AC by the wires from X and Y only, and the length l_1 is adjusted until G shows no deflection. Then the p.d. vl_1 across l_1 is balanced by the p.d. IR_1 across R_1.

Fig. 159. Comparing resistances.

$$\therefore \quad\quad\quad\quad IR_1 = vl_1.$$

Contact is then made on AC by the wires from Y and Z only, and l_2 is adjusted until G shows no deflection. In this case

$$IR_2 = vl_2.$$

Dividing
$$\frac{R_1}{R_2} = \frac{l_1}{l_2} \quad\quad\quad\quad\ldots\ldots\ldots\ldots\ldots\ldots(35.9)$$

It is important to note that at the balance point in each case the upper and lower circuits are independent of each other, since no current flows in either lead connecting them. The current in each circuit therefore remains constant throughout, and v and I do not change: therefore we are justified in cancelling them in the above equation.

Readings should always be taken in the order l_1, l_2, l_1. The second

ELECTRICITY

l_1 reading should be the same as the first; if not, one of the cells may be running down, and the readings are invalid. After (but not between) any given set of readings the rheostats H_1 and/or H_2 may be adjusted and another set obtained.

Why cannot R_1 and R_2 be placed in parallel for this experiment?

(3) Calibrating an ammeter. For this a standard cell and a standard resistance are required.

(1) *Calibrating the potentiometer wire using the standard cell.* This means determining the value of v in volt per cm along the potentiometer wire AC (Fig. 160). The standard cell of accurately known e.m.f. E_0 is selected by the two-way switch S, and the balancing length l_0 found. Then $E_0 = vl_0$,

or $$v = \frac{E_0}{l_0} \text{ in volt per cm} \dots\dots\dots\dots(35.10)$$

The potentiometer is now calibrated to read in volt, provided that v is not allowed subsequently to alter: this will happen if the cell D runs down, or if the rheostat H_1 is altered. The length l_0 should be checked frequently during the second part of the experiment to ensure that the calibration is still correct.

(2) *Using the calibrated wire to measure the p.d. across the standard resistance.* Switch S now selects the lower circuit, consisting of the standard resistance R_0, the ammeter to be calibrated, a rheostat H_2, and an accumulator D', all in series. A current I flows round this circuit, and a p.d. IR_0 therefore exists across the standard resistance.

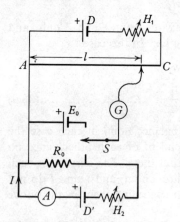

Fig. 160. Calibrating an ammeter.

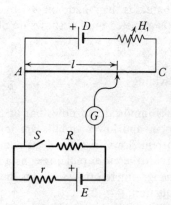

Fig. 161. Internal resistance of cell.

166

This p.d. is balanced against the p.d. across a length l of the potentiometer wire. Then $IR_0 = vl$,

or $$I = \frac{vl}{R_0} = \left(\frac{E_0}{l_0 R_0}\right)l \quad \ldots\ldots\ldots\ldots\ldots(35.11)$$

A series of values of l can be found for various values of I, which can be varied by adjusting rheostat H_2. For each value, I can be calculated from Eqn. (35.11) and the value compared with the actual ammeter reading. The error in the ammeter can be noted for values of I throughout its range.

(4) Internal resistance of a cell. The method is as described on p. 162, except that the p.d.s are measured by potentiometer instead of voltmeter. If l_0 is the balancing length with switch S open (Fig. 161) then e.m.f. of cell $E = vl_0$. If l is the balancing length with S closed and an external resistance R in the circuit, then p.d. across R is $V = vl$. Substituting these values in Eqn. (35.4),

$$\frac{l_0}{l} = \frac{R + r}{R} \quad \ldots\ldots\ldots\ldots\ldots\ldots(35.12)$$

from which the internal resistance r is calculated.

(5) Measuring small e.m.f.s. A potentiometer can be adapted to measure small e.m.f.s (e.g. thermoelectric e.m.f.s) by placing in the circuit a *large* resistance R, so as to reduce the p.d. across AC to a value comparable with that across $A'B'$ to be measured. The hot and cold junctions develop a small e.m.f. across $A'B'$, to which the potentiometer probes are connected (Fig. 162). In this experiment a very sensitive galvanometer G is necessary, and this must be protected by a high series resistance until the approximate balance point has been found.

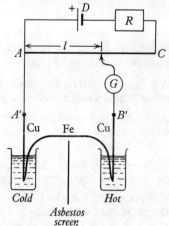

Fig. 162. Measuring thermoelectric e.m.f.s.

A numerical example illustrates how R can be calculated. Suppose the maximum p.d. to be measured is 1 millivolt (mV), where D is a 2-volt accumulator and AC a metre wire of total resistance 2 ohm.

Applying Ohm's law to the complete circuit, $2 = (2 + R)I$,

and to the wire AC only, $0.001 = 2I$.

Solving, $R = 3998$ ohm and $I = 0.5$ mA.

The potentiometer is now adapted to measure p.d.s of up to 0.001 volt, utilizing the whole of the metre wire.

Advantages of the Potentiometer. For accurate work the potentiometer has distinct advantages over the ordinary voltmeter:—
(1) It takes no current from the circuit it is measuring; it will therefore indicate the true e.m.f. of a cell, and is not liable to the errors described on p. 161.
(2) The scale is much longer—1 metre (or any other length) instead of a few centimetre.
(3) It can easily be adapted to measure any range of p.d.s.

36. Magnetism and Electromagnetism

Lines of force. A magnetic field exists round a conductor carrying an electric current; also round a magnet, since the elementary magnetic dipoles of a magnetic material are due to molecular currents. The lines of force can be demonstrated by iron filings.

A MAGNETIC FIELD LINE or LINE OF FORCE is defined as the path along which a free N pole would travel in the magnetic field.

Alternatively: the direction of the line at any point is the direction into which a compass needle, freely suspended at its centre of gravity, would turn if placed there.

The lines of force of a magnet run outwards from a N pole and inwards to a S pole. The N, or North-seeking, or positive, pole of a compass needle in the Earth's field points towards magnetic North; therefore the lines of force of the horizontal component of the Earth's field are parallel lines running from South to North.

Terrestrial magnetism. The Earth's field at a point is completely described if three quantities, called the magnetic elements, are known:

The MAGNETIC ELEMENTS are the horizontal component; the magnetic variation; and the angle of dip.

The MAGNETIC VARIATION at a point is the angle between the magnetic and geographical meridians. It is termed 'E' or 'W' according as the magnetic meridian falls to the East or West of the geographical meridian

The MAGNETIC MERIDIAN at a point is the vertical plane containing the direction of the Earth's magnetic field.

The ANGLE OF DIP is the angle between the horizontal and the direction of the Earth's total magnetic field.

A NEUTRAL POINT in a magnetic field is a point at which the net field strength is zero.

The magnetic effect of a current

For a straight wire, the magnetic field lines (lines of force) are concentric circles. For a coil, whether a flat coil or long solenoid, they form a pattern similar to that produced by a bar magnet. A coil is therefore often considered as equivalent to a bar magnet placed along its axis.

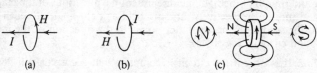

(a)   (b)   (c)

Fig. 163. Magnetic effect of a current.

(A) *Straight wire.* RIGHT HAND (CORKSCREW) RULE. If the wire is clasped in the right hand so that the thumb lies along the wire in the direction of the current, the fingers curl round the wire in the direction of the lines of force (Fig. 163a).

(B) *Coil.* If the axis of the coil is clasped in the right hand so that the thumb lies along the axis in the direction of the lines of force, the fingers curl round the axis in the direction of the current (Fig. 163b).
Alternatively: use the 'NS' rule (Fig. 163c).

The motor effect

MOTOR EFFECT: A conductor carrying a current, when placed in an external magnetic field, experiences a mechanical force.

(A) *Straight wire.* FLEMING'S LEFT HAND MOTOR RULE. If the first and second fingers of the left hand represent the directions of the magnetic flux and current, respectively, and are held at right angles to each other and to the thumb, the latter points in the direction of the force experienced by the current-bearing conductor (Fig. 164a).

(B) *Coil.* A coil experiences a torque, and the above rule may be applied.
Alternatively: consider the rotation of the 'equivalent magnet'.

169

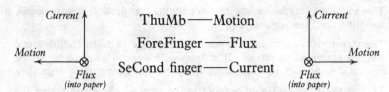

(a) Left hand motor rule.   (b) Right hand generator rule

Fig. 164. Motor and generator effects.

The generator effect

LAWS OF ELECTROMAGNETIC INDUCTION: (1) *Neumann's law* states that when a conductor cuts magnetic flux, or when the flux linking a circuit changes, an e.m.f. is induced in the conductor, or circuit. The magnitude of the induced e.m.f. is proportional to the rate of cutting, or rate of change, of the magnetic flux.

(2) *Lenz's law* states that the direction of the induced current is such as to oppose the motion, or change, causing it.

(A) *Straight wire.* FLEMING'S RIGHT HAND GENERATOR RULE. If the first finger and thumb of the right hand represent the directions of the magnetic flux and motion, respectively, and are held at right angles to each other and to the second finger, the latter points in the direction of the induced e.m.f. and current (Fig. 164*b*).

(B) *Coil.* It is only if the circuit is complete that a current will flow. Its direction is obtained by applying Lenz's law.

Quantitative electromagnetism

The following explanatory notes should be studied in conjunction with the formal theory.

In developing a system of units in electromagnetism and electrostatics, an experimental definition of one fundamental electrical unit is required, in addition to the three mechanical units, the metre, kilogramme and second. For this the ampere is chosen. It is convenient, however, to delay its formal definition, in terms of force between currents, until a few basic formulae have been developed (p. 177).

It is also possible to choose between a 'rationalized' and an 'unrationalized' system. In an unrationalized system the factor of 4π appears in formulae in which spherical symmetry is absent, and is absent from formulae possessing spherical symmetry. This state of affairs is reversed in the rationalized system, which is now used, by the insertion of 4π in Eqn. (36.6) which also gives to the value of μ_0 a factor of 4π in addition to the 10^{-7} arising from the change in size of the basic units of mass and length from g and cm to kg and m.

Lines of force and tubes of flux. In a magnetic field, the closer together the lines of force, the stronger the field. To put this another way, if we consider a bundle of lines of force, calling this a 'tube' of magnetic flux, the cross-section at points along such a tube varies inversely as the field strength.

These tubes of magnetic flux are associated with the vector B, the flux density. The unit tube of flux is such that B tubes cross unit area at a point. The total flux crossing an area A normal to the flux is thus BA.

The tubes of flux form closed loops round a conductor carrying a current, and the flux is said to 'link' the circuit. If BA tubes link a circuit having N turn, the flux linkage (Φ) of the circuit is N times as much,

or $$\Phi = BAN \dots\dots\dots(36.1)$$

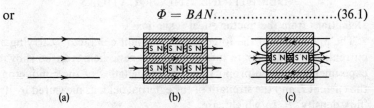

(a)   (b)   (c)

Fig. 165. Increase in flux produced by magnetic material, and (c) demagnetizing effect by the free poles formed at its ends.

The continuity of the tubes of flux is unbroken when they pass into a magnetic material. A magnetic material placed in an external magnetic field becomes magnetized by induction, and the induced magnetism produces additional tubes of flux (Fig. 165b). In the case of ferromagnetic materials this effect may increase the flux density many hundreds of times. A few of the induced tubes, however, pass back through the specimen (from R. to L., Fig. 165c) between the poles formed at its ends, reducing the total net flux from L. to R. In a long, thin, specimen these poles are far apart and the demagnetizing effect is small. A specimen in the form of a closed ring, in which there are no free poles, is the ideal arrangement for obtaining maximum flux. Clearly, the introduction of an air gap reduces its effectiveness.

Tubes of flux and the three effects. The tubes of flux are *caused* by the presence of an arrangement of electric currents in the vicinity, which is expressed by the 'magnetic effect' Eqn. (36.6).

The tubes of flux are *detected* by their two effects, the 'generator effect' and 'motor effect', which are expressed respectively by Eqns. (36.5) and (36.3).

171

ELECTRICITY

Electrical units and the speed of light. It is found by experiment that the values of μ_0 and ε_0 are related in such a way that

$$\frac{1}{\mu_0\varepsilon_0} = c^2 \qquad\qquad (36.2)$$

where c is equal to the speed of light in vacuo. The result was justified on theoretical grounds by Maxwell, who concluded that light was an electromagnetic radiation. Note that, whereas μ_0 is fixed at $4\pi \times 10^{-7}$ H m^{-1}, ε_0 is an experimental quantity, having a value of 8.85×10^{-12} F m^{-1}. This gives for c an experimental value of 3.00×10^8 m s^{-1} (see p. 193).

DEFINITIONS AND EQUATIONS

Definitions from the 'motor effect' equation

The magnitude of the force F on a straight conductor, carrying a current, perpendicular to an external magnetic field, can be shown experimentally to be proportional to the length l of the conductor, the current I, and the strength of the external field, as measured by its 'flux density' B. In which case,

$$F \propto BIl.$$

Putting the constant of proportionality equal to unity defines B.

The MAGNETIC FLUX DENSITY (B) at a point is defined by the 'motor effect' equation

$$F = BIl \qquad\qquad (36.3)$$

where F is the force on a straight conductor, length l, carrying a current I, perpendicular to an external magnetic field of flux density B. It is a vector quantity. *Unit:* weber per square metre, or tesla.

The direction of the resulting force F is given by Fleming's L.H. motor rule (p. 169).

The MAGNETIC FLUX LINKAGE (Φ) of a circuit of N turn enclosing an area A normal to a uniform magnetic field of flux density B is given by

$$\Phi = BAN \qquad\qquad (36.4)$$

Unit: weber (turn).

Derivation of the 'generator effect' equation

Two parallel conducting rails a distance apart l are short-circuited at one end and the circuit is completed by a rod lying perpendicularly

172

across the rails (Fig. 166). A uniform field of flux density B acts perpendicular to the plane of the rails. The rod is moved a distance δs to the right in a time δt.

By Fleming's R.H. generator rule an e.m.f. E and a current I are induced in the circuit in the direction shown. By Fleming's L.H. motor rule the rod experiences a force F to the left, thus opposing the motion causing it.

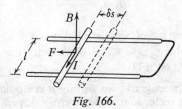

Fig. 166.

Work done pushing rod $= -F\delta s$. The minus sign occurs because the applied force is in the opposite direction to the vector F, that is, $-F$.

This energy appears as heat as the current flows through the resistance of the circuit. Heat produced $= EI\delta t$.

$$\therefore \qquad -F\delta s = EI\delta t.$$

But $F = BIl$ (Eqn. 36.3). Substituting for F we obtain

$$E = -Bl\frac{ds}{dt} = -\frac{d\Phi}{dt},$$

since $-Bl\delta s$ is the decrease in flux $-\delta\Phi$ caused by the movement of the rod.

$$\therefore \qquad E = -\frac{d\Phi}{dt} \quad\text{.............................(36.5)}$$

Thus the e.m.f. generated is *numerically equal* to the rate of cutting of flux, or rate of change of flux linkage of the circuit. The minus sign indicates Lenz's law—that the induced e.m.f. tends to oppose the change of flux causing it. The sign is of significance only when the direction of the induced e.m.f. has importance in a particular problem—e.g. in a circuit which has two e.m.f.s, one of them an induced e.m.f., acting.

173

ELECTRICITY

The 'magnetic effect' equation

The magnitude of the flux density δB at a point P due to an element length δl carrying a current I might reasonably be expected to be proportional to each of these quantities δl and I, to the sine of the angle θ as shown (Fig. 167), and inversely proportional to the square of the distance r. Thus

Fig. 167.

$$\delta B \propto \frac{I\delta l \sin \theta}{r^2}.$$

Putting the constant of proportionality equal to $\mu_0/4\pi$ (in the case of a vacuum), where $\mu_0 = 4\pi \times 10^{-7}$ henry per metre, effectively defines the *ampere* (p. 177).

LAPLACE'S RULE. The magnetic flux density B at a point in a vacuum may be calculated as the vector sum of the contributions δB from all the current elements $I\delta l$ of the circuit, where

$$\delta B = \frac{\mu_0 I\delta l \sin \theta}{4\pi r^2} \qquad \ldots\ldots\ldots\ldots\ldots\ldots\ldots(36.6)$$

the vector δB at P acting perpendicular to the plane containing δl and r, in a direction given by the corkscrew rule (Fig. 167).

This equation may thus be used to calculate the flux density B at any point for any given circuit arrangement. Although the equation cannot be verified directly for an isolated current element, indirect verification for any actual circuit is possible by comparing the calculated result with the flux density measured experimentally.

The constant μ_0 is called the PERMEABILITY OF FREE SPACE and is assigned the value of

$$\mu_0 = 4\pi \times 10^{-7} \text{ henry per metre.}$$

The 4π arises from the use of a 'rationalized' system (p. 170) and the 10^{-7} from the change from g and cm in the original units to kg and m in SI units.

The unit for μ_0 is deduced from Eqn. (36.6),

$$\frac{\text{Wb m}^2}{\text{m}^2 \text{A m}} \quad \text{or} \quad \frac{\text{Wb}}{\text{A m}} \quad \text{or} \quad \text{H m}^{-1}.$$

The PERMEABILITY (μ) of a medium is the quantity replacing μ_0 in the Laplace equation when the circuit is not in a vacuum or air. *Unit:* henry per metre.

The RELATIVE PERMEABILITY (μ_r) of a medium is the ratio of the permeability of the medium to that of free space. It is dimensionless.

174

Thus $\qquad\qquad\qquad \mu_r = \dfrac{\mu}{\mu_0}$ ....(36.7)

The following are examples of the integration of the Laplace equation. All apply to circuits in air.

B at centre of flat circular coil. The flux density B at the centre of a coil of radius r and N turn, carrying a current I, is

$$\int \frac{\mu_0 I dl \sin \theta}{4\pi r^2} = \frac{\mu_0 I}{4\pi r^2} \int dl = \frac{\mu_0 I}{4\pi r^2} 2\pi r N, \text{ since } \theta = 90°$$

and r is constant for all elements δl.

$$\therefore \qquad\qquad B = \frac{\mu_0 N I}{2r}$$ ....(36.8)

B along axis of flat circular coil. The vector δB at P due to the element δl is $\dfrac{\mu_0 I \delta l \sin \theta}{4\pi r^2} = \dfrac{\mu_0 I \delta l}{4\pi r^2}$ $(\theta = 90°)$ in direction PS (Fig. 168). This can be split into two components in directions PQ, PR. The components perpendicular to PR, from all the elements δl, cancel out, but the total component along PR, for a coil of radius a and N turn, is

$$B = \cos \phi \int \frac{\mu_0 I dl}{4\pi r^2} = \cos \phi \frac{\mu_0 I}{4\pi r^2} \int dl = \frac{a}{r} \cdot \frac{\mu_0 I}{4\pi r^2} 2\pi N a.$$

Substituting for r, where $r^2 = (a^2 + x^2)$,

$$B = \frac{\mu_0 N I a^2}{2(a^2 + x^2)^{3/2}}$$ ....(36.9)

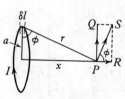

Fig. 168.

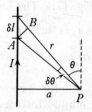

Fig. 169.

B due to a long straight wire. The vector δB at P due to the element δl is $\dfrac{\mu_0 I \delta l \sin \theta}{4\pi r^2} = \dfrac{\mu_0 I r \delta \theta}{4\pi r^2} = \dfrac{\mu_0 I \sin \theta \delta \theta}{4\pi a}$, where $\delta l \sin \theta = AB = r\delta\theta$, and $r = a/\sin \theta$ (Fig. 169).

For an infinitely long wire the flux density at P is thus

$$B = \frac{\mu_0 I}{4\pi a} \int_{\theta=0}^{\pi} \sin\theta \, d\theta = \frac{\mu_0 I}{4\pi a}\left[-\cos\theta \right]_0^{\pi} = \frac{\mu_0 I}{4\pi a}\left[1 - (-1) \right],$$

or
$$B = \frac{\mu_0 I}{2\pi a} \qquad \qquad \text{.............................(36.10)}$$

B inside a long solenoid. Within a long solenoid, of length L and N turn, carrying a current I in each turn, it can be shown that the field is almost uniform, and is given by

$$B = \frac{\mu_0 N I}{L} \qquad \qquad \text{.............................(36.11)}$$

Torque on a rectangular coil. A plane rectangular coil carrying a current I as shown lies with its plane making an angle θ with the external field of flux density B (Fig. 170). The forces acting on BC, AD cancel out. The forces on AB, CD, each of magnitude $F = BIl = BIa$, produce a torque $BIa \cdot b \cos\theta$ causing the coil to rotate about the axis EF.

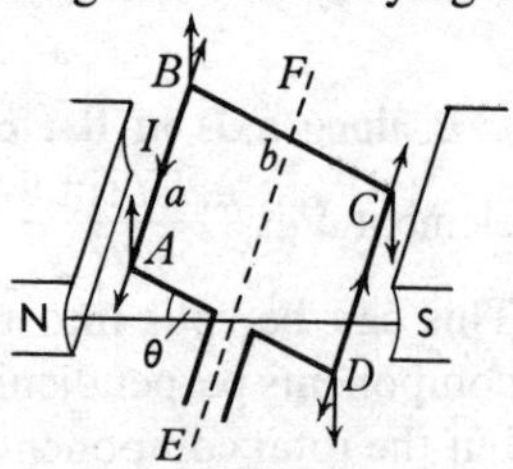

But $ab = $ Area A of coil, and if the coil has N turn,

Fig. 170.

$$\text{Torque} = BANI \cos\theta \text{......................(36.12)}$$

This expression holds for a coil of any shape.

Force between parallel straight currents. The flux density B at wire Q due to an infinite length of wire P carrying a current I_1 (in air) is $\frac{\mu_0 I_1}{2\pi a}$ (Fig. 171). The force F on a length l of the wire Q carrying a current I_2 is therefore

$$F = BI_2 l = \frac{\mu_0 I_1 I_2 l}{2\pi a},$$

substituting for B above.

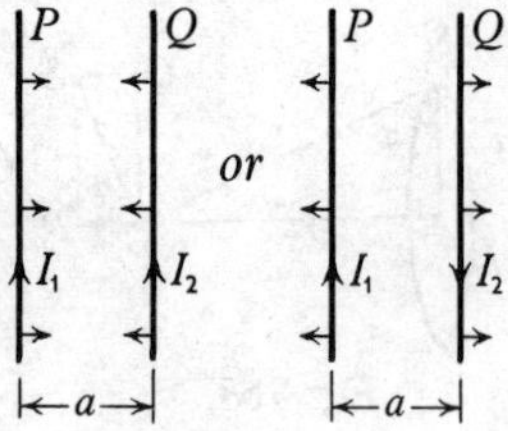

The mutual force *per unit length* between two infinite wires, a distance apart a, in vacuo or air, is therefore

Fig. 171.

$$\frac{F}{l} = \frac{\mu_0 I_1 I_2}{2\pi a} \qquad \text{.............................(36.13)}$$

'Like' currents attract, 'unlike' currents repel (Fig. 171).

36 . MAGNETISM AND ELECTROMAGNETISM

Substituting the value $\mu_0 = 4\pi \times 10^{-7}$ H m^{-1}, in the case where numerically I_1, I_2, a and l are all unity, gives a value for F of 2×10^{-7} N. This case is used to define the ampere:

The AMPERE is that constant current which, flowing through two parallel, straight, infinitely long conductors 1 metre apart, in vacuo, produces between them a mutual force of 2×10^{-7} newton per metre of their length.

Inductance

The SELF INDUCTANCE (L) of a circuit is the flux linkage of the circuit per unit current flowing in it.

The MUTUAL INDUCTANCE (M) of two coupled circuits is the flux linkage of one circuit per unit current flowing in the other.
Unit: weber (turn) per ampere, or henry.

Thus
$$\Phi = LI \quad\quad\quad\quad\quad (36.14)$$

and
$$\Phi_1 = MI_2 \quad \text{or} \quad \Phi_2 = MI_1 \quad\quad (36.15)$$

If the flux linkage is changing, an e.m.f. $E = d\Phi/dt$ is induced. Differentiating both sides w.r.t. time t,

$$E = L\frac{dI}{dt} \quad\quad\quad\quad\quad (36.16)$$

and
$$E_1 = M\frac{dI_2}{dt} \quad \text{or} \quad E_2 = M\frac{dI_1}{dt} \quad\quad (36.17)$$

Note that if iron is present in the inductor, Φ is not proportional to I, as the permeability μ is not constant. In which case, L and M vary with current in Eqns. (36.14) and (36.15). If iron is present, therefore, L and M should be defined, for a given value of current, by Eqns. (36.16) and (36.17).

37. Measurement of Flux

Theory of moving-coil galvanometer. In a *uniform field* (Fig. 172a) of flux density B the moving coil experiences a a magnetic torque $BANI \cos \theta$, as shown, when a current I flows through it. This is balanced by a torsional torque $c\theta$ in the suspension, where c is the restoring torque per unit angle of twist.

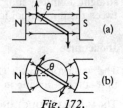

(a)

(b)

Fig. 172.

Thus $$BANI \cos \theta = c\theta,$$

resulting in a non-linear scale (θ not proportional to I).

In a *radial field* (Fig. 172b), provided by curved pole-pieces and soft-iron core, the plane of the moving coil is always in the field direction. The magnetic torque is thus $BANI$ for all angles θ. The torsional torque is $c\theta$ as before. Therefore $BANI = c\theta$, and the scale is linear:

$$I = \left(\frac{c}{BAN} \right) \theta \dots\dots\dots\dots\dots(37.1)$$

where θ is the steady deflection.

Ballistic galvanometer. If the suspension of a moving-coil galvanometer has a large moment of inertia and a weak restoring torsion—as shown by a large period of swing, and if the coil is wound on a non-metallic frame so that there is a minimum of electromagnetic damping, then the instrument can be used ballistically as well as for measuring steady currents. Theory shows that the ballistic 'throw' is a measure, not of the current, but of the total charge passing. Provided this passes almost instantaneously, θ is proportional to Q,

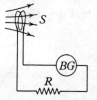

or $$Q = k\theta \dots\dots\dots(37.2)$$

where θ is the *ballistic throw*.

The ballistic galvanometer can be used as a *fluxmeter.* In Fig. 173 let the flux linkage of the circuit of total resistance R be changing at a rate $d\Phi/dt$. At any instant during the change the induced e.m.f. $E = \dfrac{d\Phi}{dt}$, and the induced current I is given by $E = IR$.

Fig. 173. Ballistic galvanometer for flux measurement.

$$\therefore \qquad \delta\Phi = E\delta t = IR\delta t = R\delta Q,$$

where $I\delta t = \delta Q$, the charge passing in the time δt, during which time the flux linkage changes by $\delta\Phi$. Integrating for a total *change* of flux linkage Φ resulting in a flow of charge Q round the circuit,

$$\Phi = RQ \quad ....(37.3)$$

If this causes a ballistic throw θ, substituting from Eqn. (37.2),

$$\Phi = kR . \theta \quad ....(37.4)$$

Ballistic method for measuring flux density B. A small search coil S (N turn, area A) is connected in series with a ballistic galvanometer BG in a circuit of total resistance R (Fig. 173). S is placed in the magnetic field of unknown flux density B, with the plane of the coil normal to the flux. The flux linkage Φ is thus BAN.

The current in the primary circuit producing the flux is then switched off. Alternatively, if the flux is concentrated across a small air gap between the poles of a magnet, the search coil is rapidly withdrawn from the field. In either case, the flux linkage changes from Φ to zero. A charge Q flows through the galvanometer, causing a ballistic throw θ. From Eqn. (37.4),

$$BAN = \Phi = kR . \theta ....(37.5)$$

To determine the constant (kR) the galvanometer may be connected to the secondary of a known standard mutual inductance M (Fig. 174). The throw θ' is noted when a known current I is switched off in the primary circuit; the change of flux Φ' causing this throw is calculated from $\Phi' = MI$.

Mutual inductance of concentric solenoids in air. The secondary coil (N_2 turn) is wound round the middle of a long primary coil (N_1 turn, area A, length L). For this arrangement the value of M is easily calculated, and it may therefore be used as a standard (Fig. 174).

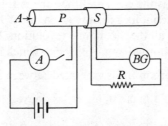

Fig. 174. Calibrating fluxmeter using standard mutual inductance.

The flux linkage Φ of the secondary due to current I in the primary is given by $\Phi = BAN_2 = \dfrac{\mu_0 N_1 I}{L} AN_2$. But $\Phi = MI$,

$$\therefore \qquad M = \frac{\mu_0 N_1 N_2 A}{L}.$$

Steady current method for measuring flux density B. A circular conducting disc of area A is rotated about its centre in a plane perpendicular to a magnetic field of flux density B at a constant speed in rev s^{-1}. A radius of the disc thus sweeps out an area per second An, and cuts flux at a rate BAn. A steady e.m.f. $E = \dfrac{d\Phi}{dt} = BAn$ is generated between the centre and the periphery. Contact is made at these two points by brushes and the circuit is completed by a microammeter which measures the current I generated (Fig. 175a). Since $E = IR$, where R is the total resistance of the circuit,

$$IR = BAn \dotfill (37.6)$$

hence B.

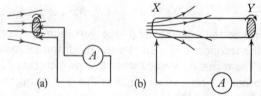

Fig. 175. Rotating-disc, and rotating-tube, methods for flux measurement.

An alternative method, suitable if the flux is produced within a long solenoid, is to use a long hollow conducting tube, placed coaxially with the solenoid and in such a position that effectively all the flux entering the tube through the end X leaves the tube through the sides (Fig. 175b). If this is the case, and the tube is rotated at a constant speed n in rev s^{-1} about its axis, any given line XY along the outside of the tube will cut all the flux in one revolution. If B is the uniform flux density over a normal area A through which all this flux passes, then $\dfrac{d\Phi}{dt} = BAn$. An e.m.f. E is generated between X and Y and drives a current I through a microammeter, as above.

The Earth inductor. The coil (N turn, area A) of the Earth inductor is connected in series with a ballistic galvanometer in a circuit of total resistance R, and placed with its plane vertical and in the magnetic E–W direction. When rotated quickly through 180° about the vertical diameter, the flux through the coil is reversed, and the change of flux linkage is therefore $2B'AN$, where B' is the flux density of the horizontal component of the Earth's field. The ballistic throw θ' is given by

$$2B'AN = \Phi' = kR . \theta'.$$

When placed in a horizontal plane and rotated through 180° about the horizontal diameter in the magnetic meridian, the ballistic throw θ'' is given by

$$2B''AN = \Phi'' = kR . \theta'',$$

where B'' is the flux density of the vertical component.

Hence the angle of dip a is given by

$$\tan a = \frac{B''}{B'} = \frac{\theta''}{\theta'},$$

and may be determined using the galvanometer uncalibrated. To determine absolute values of B' and B'' the galvanometer must be calibrated as a fluxmeter, as previously described.

The Earth inductor, if provided with a split slip-ring used as a commutator, can be rotated at a steady rate n in rev s^{-1} and will generate a mean e.m.f. of

$$E = \frac{d\Phi}{dt} = 4BANn,$$

and a mean current $I = \dfrac{4BANn}{R}$. Hence B', B'' and a.

Earth's horizontal field by tangent galvanometer. The coil (N turn, radius r) of the tangent galvanometer is placed with its plane in the magnetic meridian, as shown by the compass needle at its centre (Fig. 176). The circuit consists of a reversing switch S, rheostat,

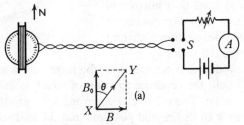

Fig. 176. Tangent galvanometer.

181

ammeter and low voltage d.c. supply. The coil must be well away from any external magnetic influences, including those of the rest of the circuit.

If a current I is now passed this will result in a deflection θ of the compass needle, since the needle will simply take up the direction XY of the resulting field at the centre of the coil (Fig. 176a). This field is now the vector resultant of two fields at right angles—the Earth's horizontal field B_0 and the field B produced by the current in the coil. Therefore the deflection θ is given by

$$B = B_0 \tan \theta \quad \dots\dots\dots\dots\dots\dots(37.7)$$

where $B = \dfrac{\mu_0 NI}{2r}$. Hence B_0.

Values of I against θ can be obtained by adjusting the rheostat; the deflection θ should be restricted to the range 30° to 70°. For each value of I, readings θ of both ends of the pointer should be taken, and again with the current reversed; the mean value θ of four readings is thus obtained.

Absolute measurements

In order to make an absolute determination of an electrical quantity, it is necessary to use a method of measurement based directly on the fundamental principles used in its definition, without any reference to previously calibrated instruments. If the ampere and ohm are determined independently in this way, the other units readily follow.

Determination of current by current balance. By the apparatus shown (Fig. 177), or any similar arrangement, current is measured by a direct application of first principles. A small coil (N_2 turn, area A) pivoted at XY is situated at the centre of a large, flat coil (N_1 turn, radius r) fixed in a horizontal plane. The two coils are connected in series and the unknown current I is passed through them.

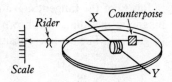

Fig. 177. Current balance

The field produced at the centre of the large coil causes a torque on the small coil, the resulting rotation of which is indicated by a pointer on a scale. The rider, of known mass, is adjusted to bring the pointer back to its original position, and the magnetic torque T exerted is thus calculated from the movement of the rider.

Now $T = BAN_2I$, where B at centre of large coil $= \dfrac{\mu_0 N_1 I}{2r}$.

$$\therefore \qquad I = \sqrt{\dfrac{2Tr}{\mu_0 N_1 N_2 A}}.$$

Lorenz determination of resistance. The circuit (Fig. 178) is a potentiometer arrangement by which the p.d. *IR* across the unknown resistance R can be balanced against the e.m.f. E generated by a disc (area A) rotating in a magnetic field inside a long solenoid (N turn, length L). The current I flows through both resistance and magnetizing solenoid: it therefore affects both E and IR equally, and is eliminated in the final expression for R. R is thus measured directly from first principles, without the assistance of calibrated electrical instruments.

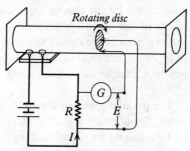

Fig. 178. Lorenz determination of resistance.

The speed of rotation n (in rev s $^{-1}$) of the disc is adjusted until the galvanometer G shows no deflection, in which case $E = IR$.

But
$$E = \dfrac{d\Phi}{dt} = BAn = \dfrac{\mu_0 NI}{L}An.$$

$$\therefore \qquad R = \dfrac{\mu_0 NAn}{L}.$$

38. Electrostatics

A preliminary revision of the simple qualitative phenomena of electrostatics, including induced charges, the gold leaf electroscope, the electrophorus, etc., is assumed in this Section.

Faraday's ice pail experiment. When an insulated conductor, charged with $+Q$, is lowered into a deep metal can connected to an electroscope, a charge $-q$ is induced on the inside of the can—the corresponding charge $+q$ being repelled to the outside of the can, and to the leaves of the electroscope, which consequently diverge (Fig. 179a). After it has made contact with the can, the conductor is removed and is, found to be totally discharged (Fig. 179b). The divergence of the leaves remains unchanged throughout.

As the conductor is totally discharged, its own charge $+Q$ must have been exactly neutralized by the charge $-q$ induced on the inside of the can. In other words $Q = q$, and the experiment illustrates the principle that a *charged conductor always induces an equal and opposite charge on neighbouring conductors.*

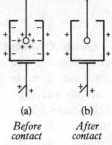

(a)   (b)

Before   *After*
contact   *contact*

Fig. 179. Faraday's ice pail experiment.

The parallel plate condenser. Suppose we have an insulated metal plate A, on which a charge $+Q$ has been placed. By the principle just explained an equal and opposite charge $-Q$ will be induced on surrounding objects—in this case the walls, floor and ceiling, if no other conductors are present.

A second insulated, but uncharged, metal plate B is now brought up close to A (shown diagrammatically in Fig. 180a). If B is very close to A, so that other surrounding bodies can be ignored (this will never be quite true), charges $-Q$ and $+Q$ separate out, by induction, on B. If B is now earthed, the charge $+Q$ flows away, leaving B negatively charged (Fig. 180b).

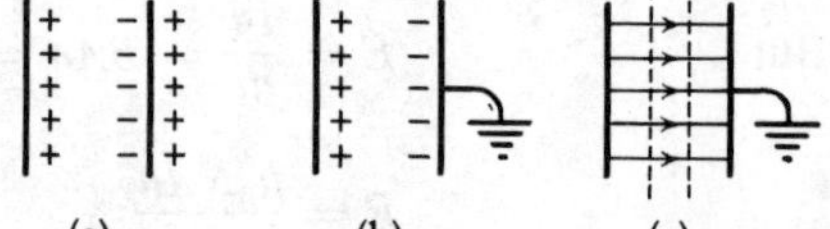

Fig. 180.

An electric field of strength E exists in the space between A and B, the lines of electric force starting, by convention, on the $+$ charge on A and ending on the corresponding $-$ charge on B (Fig. 180c). A line of electric force is defined as the path of a free $+$ charge: it therefore follows the direction of the vector E at every point, since E is the force on unit $+$ charge (see definitions). In the present case, however, the charges are unable to move, as the space is a 'dielectric', or insulator; but if a conducting wire is connected between the plates the charges do flow together and neutralize one another, discharging the condenser.

Work is done moving a $+$ charge from B to A, against the electric field *E*, and therefore a p.d. *V* exists between the plates. 'Lines of equipotential' are shown, as broken lines, in Fig. 180c. Using our analogy between potential and 'vertical height' (p. 158) we see that the lines of equipotential correspond to 'contour lines' on a map— the + charge tending to flow 'down hill', in the direction of *E*, which is the maximum negative potential gradient (see definitions again) and is always at right angles to the equipotential lines. It should be noted, however, that *E* is a *vector* quantity but *V* is a *scalar*.

The lines of electric force can be regarded as producing tubes of electric flux (note the similarity to magnetism). A tube of flux begins on a + charge and ends on an equal − charge induced on a neighbouring conductor.

Effect of a dielectric. An air condenser is charged to a p.d. *V* by connecting a battery across the plates (Fig. 181a). With the battery still connected, let a dielectric medium now be introduced into the air space, to fill it completely. The medium becomes 'polarized' by induction, the + and − charge of each molecule separating to form an electric dipole (Fig. 181b). A negative charge thus appears on the surface of the dielectric facing the + plate, and a positive charge on that facing the − plate (analogous to the induced poles of a magnet). Additional tubes of flux appear on account of this induced charge, and the charge *Q* on the plates is correspondingly increased (shown diagrammatically in Fig. 181c), the p.d. *V* being kept constant by the battery. The capacitance *Q/V* of the condenser is thus increased by the introduction of the dielectric.

If, instead, the battery in Fig. 181a is removed after charging, leaving the plates insulated, the subsequent insertion of the dielectric cannot alter the charge *Q* on the plates. In this case the p.d. *V* is reduced, since the induced charge on the dielectric partially neutralizes the charge on the plates and renders the electric field less intense (Fig. 181d). Again, the capacitance *Q/V* is increased, in the same ratio as before.

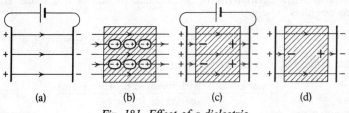

Fig. 181. Effect of a dielectric.

DEFINITIONS AND EQUATIONS

The inverse square law in electrostatics

The magnitude of the mutual force F between two point charges Q_1 and Q_2, in vacuo, can be shown experimentally to be proportional to the product of the charges and inversely proportional to the square of their distance apart r. Thus

$$F \propto \frac{Q_1 Q_2}{r^2}.$$

Putting the constant of proportionality equal to $1/4\pi\varepsilon_0$ (in the case of a vacuum) defines ε_0, the *permittivity of free space*.

The PERMITTIVITY OF FREE SPACE (ε_0) is defined by the equation

$$F = \frac{Q_1 Q_2}{4\pi\varepsilon_0 r^2} \quad\dots\dots\dots\dots\dots\dots\dots\dots\dots\dots(38.1)$$

where F is the mutual force between two point charges Q_1, Q_2 a distance apart r, in vacuo. *Unit:* farad per metre.

The 4π arises from the use of a 'rationalized' system. Since the magnitudes of the units of force F, distance r, and charge Q in coulomb, are all previously defined, the magnitude of ε_0 in Eqn. (38.1) is an *experimental* quantity, its measured value being

$$\varepsilon_0 = 8.85 \times 10^{-12} \text{ farad per metre.}$$

This is found to be equal to $1/\mu_0 c^2$ (Maxwell's relation), where $c = $ Speed of light in vacuo $= 3.00 \times 10^8$ m s^{-1}, and $\mu_0 = 4\pi \times 10^{-7}$ H m^{-1}.

The unit for ε_0 is deduced from Eqn. (38.1),

$$\frac{C^2}{N\,m^2} \quad \text{or} \quad \frac{C^2}{J\,m} \quad \text{or} \quad \frac{C}{V\,m} \quad \text{or} \quad F\,m^{-1}.$$

The PERMITTIVITY (ε) of a medium is the quantity replacing ε_0 in Eqn. (38.1) when the charges are not in a vacuum or air. *Unit:* farad per metre.

The RELATIVE PERMITTIVITY OT DIELECTRIC CONSTANT (ε_r) of a medium is the ratio of the permittivity of the medium to that of free space. It is dimensionless.

Thus

$$\varepsilon_r = \frac{\varepsilon}{\varepsilon_0} \quad\dots\dots\dots\dots\dots\dots\dots\dots\dots\dots(38.2)$$

Alternatively, ε_r may be defined from

$$\varepsilon_r = \frac{\text{Capacitance with medium between plates}}{\text{Capacitance in vacuo}} \quad\dots\dots\dots(38.3)$$

for any capacitor.

186

The ELECTRIC FIELD STRENGTH (E) at a point is the force per unit charge on a positive charge placed at that point. It is a vector. *Unit:* newton per coulomb. (Alternative definition, below.)

The force F on a charge Q placed in an electric field of strength E is therefore

$$F = QE \quad \dots\dots\dots\dots\dots\dots(38.4)$$

It follows from Eqns. (38.1) and (38.4) that the electric field E at a distance r from a charge Q, in vacuo, is given by

$$E = \frac{Q}{4\pi\varepsilon_0 r^2} \quad \dots\dots\dots\dots\dots(38.5)$$

Electric potential

The ELECTRIC POTENTIAL (V) at a point is the work done per unit charge bringing a positive charge from infinity to the point. *Unit:* joule per coulomb, or volt.

The 'potential difference' between two points is thus the work done moving unit positive charge between the points. This corresponds with the definition of V in joule per coulomb in current electricity (p. 156).

Electric field is negative potential gradient. If a charge δQ is moved through a distance δs against an electric field E, the Work done = Force $\times$ Distance $= -E\delta Q . \delta s$. The minus sign occurs because the applied force is in the opposite direction to the vector E, that is, $-E$.

Work done per unit charge $= -E\delta s$. This by definition is the potential difference δV between the two points.

$$\therefore \qquad \delta V = -E\delta s,$$

or

$$E = -\frac{dV}{ds}\dots\dots\dots\dots\dots\dots(38.6)$$

Thus electric field strength E can be alternatively defined:

The ELECTRIC FIELD STRENGTH (E) at a point is the negative potential gradient at that point. It is a vector. *Unit:* volt per metre. (Alternative definition, above.)

Electric potential due to point charge in vacuo. The potential at a point P a distance r from a point charge $+Q$ at O is the work done bringing unit positive charge from infinity to P (Fig. 182). The work done moving the unit charge a distance $-\delta s$ (i.e. inwards) at a distance s from O is

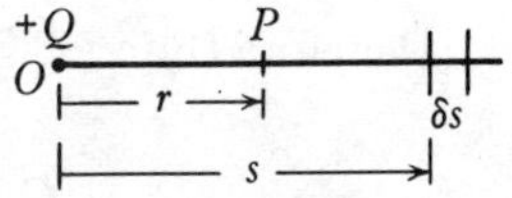

Fig. 182.

$$\delta V = \text{Force} \times \text{Distance} = E(-\delta s),$$

187

where E, or $\dfrac{Q}{4\pi\varepsilon_0 s^2}$, is the field strength at that point, due to Q, in vacuo or air.

Total work done moving the unit charge from infinity to P

$$= V = \int_{s=\infty}^{r} dV = -\int_{\infty}^{r} \frac{Q}{4\pi\varepsilon_0 s^2} ds = -\left[-\frac{Q}{4\pi\varepsilon_0 s}\right]_{\infty}^{r}$$

or
$$V = \frac{Q}{4\pi\varepsilon_0 r} \quad\dots\dots\dots\dots\dots\dots\dots(38.7)$$

which is the potential V at a distance r from a point charge Q, in vacuo or air.

N.B. Compare the inverse square law for E (Eqn. 38.5) with the inverse law for V (Eqn. 38.7).

Capacitance

The CAPACITANCE (C) of a conductor is the charge required per unit rise of potential.

The CAPACITANCE (C) of a capacitor is the charge on either plate required per unit rise of p.d. between the plates.   *Unit:* coulomb per volt, or farad.

Thus
$$C = \frac{Q}{V} \quad\dots\dots\dots\dots\dots\dots\dots(38.8)$$

Potential due to a charged spherical conductor. A charge placed on an isolated spherical conductor (whether solid or hollow) distributes itself uniformly over the surface. It can be shown that
(1) for points *outside* the sphere, the potential V is the same as if the whole of the charge were concentrated at the centre, and
(2) for points *inside* the sphere, the potential is uniform, at a value equal to that at the surface.

These facts are used in the following derivations.

Capacitance of an isolated spherical conductor. The potential V at the surface of the sphere, of radius a, is the same as if the whole of the charge $+Q$ on its surface were concentrated at the centre.

Therefore (Eqn. 38.7),
$$V = \frac{Q}{4\pi\varepsilon_0 a},$$

$$\therefore \qquad C = \frac{Q}{V} = \frac{Q}{Q/4\pi\varepsilon_0 a} = 4\pi\varepsilon_0 a \quad\dots\dots\dots\dots(38.9)$$

for a sphere in vacuo or air.

Capacitance of two concentric spheres. A charge $+Q$ is placed on the inner sphere (radius a), and the outer sphere (radius b) is earthed. A charge $-Q$ is induced on the inside of the outer sphere, and the corresponding charge $+Q$ flows to earth (Fig. 183).

Potential at A due to $+Q$ on $A = +Q/4\pi\varepsilon_0 a$ (point *outside* sphere)

,, ,, ,, ,, ,, $-Q$,, $B = -Q/4\pi\varepsilon_0 b$ (,, *inside* ,,)

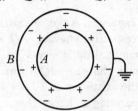

Fig. 183. Concentric spheres capacitor.

$\therefore$ Total potential at $A = \dfrac{Q}{4\pi\varepsilon_0 a} - \dfrac{Q}{4\pi\varepsilon_0 b}.$

Potential at B due to $+Q$ on $A = +Q/4\pi\varepsilon_0 b$ (point *outside* sphere)

,, ,, ,, ,, ,, $-Q$,, $B = -Q/4\pi\varepsilon_0 b$ (,, *outside* ,,)

$\therefore$ Total potential at B is zero (as it must be, since B is earthed).

$\therefore$ P.d. between the spheres $= V = \dfrac{Q}{4\pi\varepsilon_0 a} - \dfrac{Q}{4\pi\varepsilon_0 b}.$

$\therefore$
$$C = \frac{Q}{V} = \frac{4\pi\varepsilon_0 ab}{b - a} \quad\dots\dots\dots\dots\dots(38.10)$$

where the space between the spheres is a vacuum or air.

Capacitance of a parallel plate capacitor. The capacitance of two concentric spheres of *very large* radius

$$= C = \frac{4\pi\varepsilon_0 ab}{b - a} \simeq \frac{4\pi\varepsilon_0 a^2}{d} \simeq \frac{4\pi\varepsilon_0 A'}{4\pi d},$$

where   Distance between plates $= b - a = d$, where $d \ll a$ or b,

and   Area of each plate $= A' \simeq 4\pi a^2$.

A parallel plate capacitor (area of each plate A) can be regarded as a portion of the above capacitor. As capacitance is proportional to area of plate, its capacitance is given by

$$C = \frac{A}{A'} \cdot \frac{4\pi\varepsilon_0 A'}{4\pi d} = \frac{\varepsilon_0 A}{d} \quad\dots\dots\dots\dots(38.11)$$

189

ELECTRICITY

where d is the distance between the plates, separated by a vacuum or air.

With a material of permittivity ε between the plates, the capacitance is given by

$$C = \frac{\varepsilon A}{d} \quad \dots\dots\dots\dots\dots\dots(38.12)$$

Energy of a charged capacitor. A p.d. V exists across the plates, as a consequence of the respective charges $+Q$ and $-Q$ upon them. By the definition of potential difference V (p. 156) the work done separating an additional charge $+\delta Q$ from the negative to the positive plate, across the existing p.d. V, is $V\delta Q$.

Total work done separating a charge Q is

$$\int_0^Q V dQ = \int_0^Q \frac{Q}{C} dQ = \tfrac{1}{2}\frac{Q^2}{C},$$

where C is the capacitance Q/V. If V is the final p.d. across the plates,

$$\text{Energy} = \tfrac{1}{2}\frac{Q^2}{C} = \tfrac{1}{2}QV = \tfrac{1}{2}CV^2 \dots\dots\dots\dots(38.13)$$

Capacitors in series. A p.d. V is applied across two capacitances C_1, C_2 in series (Fig. 184a). A charge Q separates on to each plate as shown, and p.d.s V_1, V_2 exist across the two capacitors, where

$$V = V_1 + V_2.$$

But $\quad V = \dfrac{Q}{C}, \quad V_1 = \dfrac{Q}{C_1}, \quad V_2 = \dfrac{Q}{C_2},$ (a)

applying Eqn. (38.8) to each capacitor in turn, and to the whole, considered as a single capacitance C.

Substituting and cancelling Q,

$$\frac{1}{C} = \frac{1}{C_1} + \frac{1}{C_2} \quad \dots\dots\dots\dots(38.14)$$ (b)

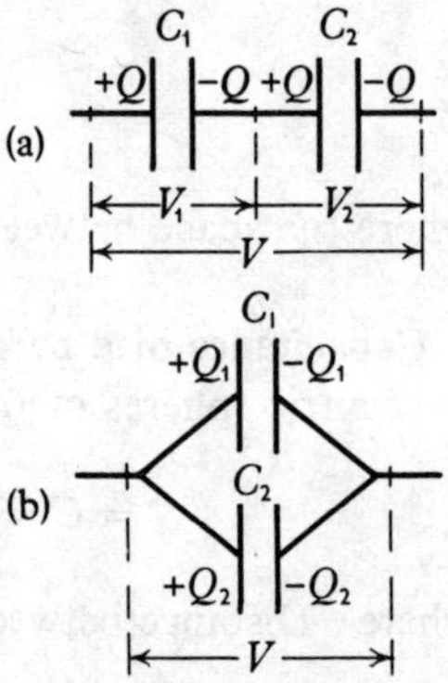

Fig. 184. *Capacitances in series and parallel.*

Capacitors in parallel. A p.d. V is applied across two capacitances C_1, C_0 in parallel (Fig. 184b). Charges Q_1, Q_2 separate on to the plates as shown, the total charge separated being Q, where

$$Q = Q_1 + Q_2.$$

But $\qquad Q = CV, \qquad\qquad Q_1 = C_1V, \qquad\qquad Q_2 = C_2V,$

190

applying Eqn. (38.8) to each capacitor in turn, and to the whole, considered as a single capacitance C.

Substituting and cancelling V,

$$C = C_1 + C_2 \quad \dots\dots\dots\dots(38.15)$$

Composite dielectrics. The parallel plate capacitor in Fig. 185 contains two dielectrics of thickness d_1, d_2 and permittivities ε_1, ε_2, respectively. The area of each plate is A. Effectively this is two capacitors in series, the total capacitance C being given by

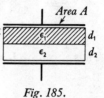

Fig. 185.

$$\frac{1}{C} = \frac{1}{C_1} + \frac{1}{C_2}, \quad \text{where} \quad C_1 = \frac{\varepsilon_1 A}{d_1} \quad \text{and} \quad C_2 = \frac{\varepsilon_2 A}{d_2}.$$

Thus
$$C = \frac{A}{\left(\dfrac{d_1}{\varepsilon_1} + \dfrac{d_2}{\varepsilon_2}\right)}.$$

39. Measurement of Capacitance

In order to measure capacitance (and hence permittivity) either Q or V (or both) must be measured, since $C = Q/V$. If capacitances are merely to be compared, and not determined absolutely, uncalibrated instruments, in which the deflection is merely proportional to the charge or p.d., can be used. The methods available can be divided into two categories: those in which Q is found by discharge through a galvanometer, and those in which V is measured by an electrostatic voltmeter, e.g. a gold leaf electroscope. The methods are outlined below, from which the principles should be clear; but many other variations are possible.

Measurement of Q by discharge. Switch to A (Fig. 186a) to charge the capacitor; switch to B to discharge through the ballistic galvanometer, which indicates a throw θ proportional to the charge Q passed. Repeat with a second capacitor substituted.

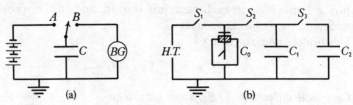

Fig. 186. Comparison of capacitances.

Thus $\qquad Q_1 = C_1 V \qquad$ and $\qquad Q_2 = C_2 V.$

$$\therefore \qquad \frac{C_1}{C_2} = \frac{Q_1}{Q_2} = \frac{\theta_1}{\theta_2} \dots\dots\dots\dots(39.1)$$

giving a means of comparing two capacitances.

Measurement of I by continuous charge and discharge. The circuit as above is used, but with a milliammeter instead of a ballistic galvanometer, and a charge-discharge switch vibrating at a rate n in oscillation per second. A current I equal to nQ thus flows through the milliammeter, which shows a steady deflection.

But $Q = CV$ for a given capacitor, so

$$I = nCV \dots\dots\dots\dots(39.2)$$

If n is known, and current I and p.d. V are measured with calibrated instruments, the capacitance C can be determined *absolutely* by this method.

Measurement of V by sharing charges. Two capacitances C_1, C_2 to be compared are connected in parallel with each other and with the gold leaf electroscope, which itself has a capacitance C_0 between the leaf and earthed case (Fig. 186b). The electroscope may be charged by an H.T. battery as shown, or by supplying a series of small charges from an electrophorus, or by the method of induction using a charged rod. Suitable insulated switches S_1, S_2, S_3 enable the charges to be given and shared as follows.

(1) The capacitance of the electroscope is first determined in terms of one of the capacitances C_1, so that it may be eliminated from the final equation.

Close switch S_1; this charges C_0 with a charge Q. The deflection of the leaf indicates a p.d. V_1 across the electroscope.

Open S_1, close S_2; the charge is now shared with C_1 in parallel. The p.d. indicates on the electroscope falls to V_2.

Thus $\qquad Q = C_0 V_1 = (C_0 + C_1) V_2.$

$\therefore \qquad \dfrac{C_0 + C_1}{C_0} = \dfrac{V_1}{V_2},$ hence C_0 in terms of C_1.

(2) Now close S_2, close S_1; this charges C_0 and C_1 in parallel, with a total charge Q'. The p.d. indicated is V_3.
Open S_1, close S_3; the charge is now shared with C_2 in parallel. The p.d. falls to V_4.

Thus $\qquad Q' = (C_0 + C_1) V_3 = (C_0 + C_1 + C_2) V_4.$

$\therefore \qquad \dfrac{C_0 + C_1 + C_3}{C_0 + C_1} = \dfrac{V_3}{V_4}.$

Hence $\dfrac{C_1}{C_2}$, substituting for C_0 and rearranging.

Measurement of permittivity. The above methods can be used with capacitors of any type. To obtain a value for the permittivity of a medium, however, particular types of capacitor are required.

(1) To measure *relative permittivity* (ε_r) or *dielectric constant*. For a liquid, compare the capacitance of a suitable capacitor with air between its plates with its capacitance when filled with the liquid. For a solid, compare the capacitance of a parallel plate air capacitor with that of a capacitor of the same dimensions having the dielectric between its plates.

Hence $\qquad \varepsilon_r = \dfrac{\text{Capacitance with substance between plates}}{\text{Capacitance in air}}.$

(2) To measure *absolute permittivity of air* (ε_0). A parallel plate air capacitor is used, of which the dimensions can be accurately measured. Its capacitance in farad is also measured with calibrated instruments, e.g. from Eqn. (39.2). From all these readings c, the speed of light in vacuo, can be calculated (see p. 172).

From the dimensions, $C = \varepsilon_0 A / d$; and by calibrated instruments, $C = Q/V$. Hence ε_0.

Hence c from $c^2 = \dfrac{1}{\mu_0 \varepsilon_0}$ (Eqn. 36.2).

40. Electrical Machines

A 'machine' takes in energy in one form and delivers it in another, or modified, form. A generator takes in mechanical, and delivers electrical, energy; a motor takes in electrical energy, and delivers mechanical. A transformer converts electrical energy from one voltage to another.

Principles of a.c. and d.c. generators. The e.m.f. generated by a coil rotating at an angular velocity ω in a uniform field is given by $E = E_0 \cos \omega t$ (Eqn. 42.1). If this is tapped from two slip-rings as in Fig. 187a, the e.m.f. obtained across AB is alternating, as shown. If a single split slip-ring is used, as in Fig. 187b, the brushes change contact to opposite leads at the same instant that the e.m.f. changes

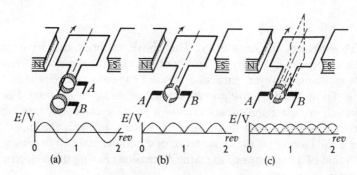

Fig. 187. A.c. and d.c. generators.

direction in the coil, and the e.m.f. across AB is therefore unidirectional, though not steady. With two coils at right angles, and four sectors in the slip-ring (Fig. 187c), a steadier e.m.f. is obtained. The larger the number of coils the steadier the e.m.f., since each coil is tapped only when delivering its peak value. This peak value occurs when the coil is passing through the position shown in Fig. 187b.

The d.c. motor. The simple d.c. generator (Fig. 187b) converts the mechanical energy supplied in turning the armature into electrical energy. Conversely, if a direct current is passed through the same apparatus the armature turns, providing mechanical energy, and it becomes a d.c. motor. The direction of rotation, however, is now

194

reversed, since Fleming's left hand motor rule, instead of the right hand generator rule, applies.

Back motor effect in a generator. A mechanical force applied to turn the armature results in the generation of an e.m.f. across its terminals. If the circuit is not completed, however, no current flows, and no electrical energy is obtained—the machine is just a coil being turned against friction.

If a load resistance is connected across the terminals a current I flows, and the armature becomes subjected, in addition to the friction, to an electromagnetic torque $BANI \cos \theta$ (Eqn. 36.12) at every instant, which also opposes the motion. This is the 'back motor effect', and increases in magnitude as the load current is increased, making the armature harder to rotate. As more electrical energy is obtained, more mechanical energy has to be supplied.

The efficiency of a large generator may be 90%. The causes of inefficiency can be classified as iron (hysteresis), copper (heating), and frictional losses.

Back generator effect in a motor. At the instant the motor is switched on a current I flows, where

$$E = IR \qquad \qquad (40.1)$$

E being the applied forward e.m.f. and R the total resistance of the circuit. As the armature accelerates, the coil cuts flux at an increasing rate, causing a back e.m.f. $E' = d\Phi/dt$ to be generated across the armature terminals. The current I diminishes as the speed of rotation increases, since now

$$E - E' = IR \qquad \qquad (40.2)$$

If no mechanical load is taken, and in the absence of friction, the speed increases until $E' = E$, and the current falls to zero. No electrical energy is being supplied, and no mechanical energy delivered.

If a mechanical load is now applied to the armature the speed drops, E' decreases, and I increases. The greater the mechanical load, the greater the current taken. In general, the power equation is

$$EI = E'I + I^2R \qquad \qquad (40.3)$$

where EI is the power supplied, $E'I$ the power converted to mechanical and frictional energy, and I^2R the power converted to heat in the circuit.

The meaning of the term 'back e.m.f.', defined on p. 157, is illustrated here. The total potential *rise*, E, applied to the circuit by the source of electrical energy, is equal to the sum of the potential *drop*, or 'back e.m.f.', E', across the armature, due to the conversion of electrical energy into mechanical energy, and the potential *drop IR*, in the circuit, due to the conversion of electrical energy into heat.

Starting resistance in a motor. The resistance R of the armature coil in a motor is very small, to minimize heat losses, and I could be very large at starting, before the back e.m.f. had built up. This current could burn out the armature. An additional starting resistance is therefore incorporated, which is gradually reduced as the speed increases (Fig. 188).

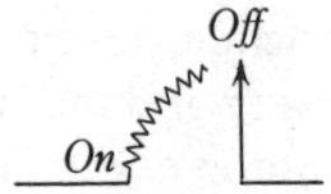

Off

On

Fig. 188.

Field coils. Except in small dynamos, the field in which the armature rotates is provided by an electromagnet. The field current may be derived from the same circuit as the armature current, the field coils being in series, or parallel, with the armature coils, or a compound of both types of winding (Fig. 189). A series-wound motor

Armature

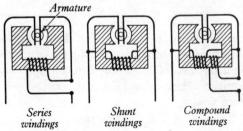

| Series windings | Shunt windings | Compound windings |

Fig. 189. Field coils in generators and motors.

gives a large initial torque on starting, and is therefore suitable when good acceleration is required, e.g. in Underground trains. A shunt-wound motor gives a more uniform torque, and is suitable when the speed should be independent of the load, e.g. in lifts. A generator, to give an e.m.f. independent of the current taken, requires a small fraction of series, and a large fraction of shunt, windings.

A.c. transformer. Coils A (few turn N_1) and B (many turn N_2) are wound on a laminated soft-iron core, so that almost all the flux produced by one circuit passes through the other (Fig. 190). If an alternating current is applied to one coil (the primary) an alternating e.m.f. is induced across the other (the secondary). The current in the primary, and therefore the flux Φ linking the secondary, is changing at the maximum rate at Q and S, and the induced e.m.f. $d\Phi/dt$ across the secondary is maximum at these instants, as shown. At P and R the rate of change is zero, and the induced e.m.f. is zero.

It can be shown that, to a close approximation,

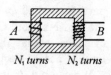

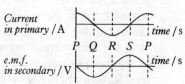

Fig. 190. A.c. transformer.

$$\frac{\text{Primary voltage } E_1}{\text{Secondary voltage } E_2} = \frac{N_1}{N_2} \quad ....(40.4)$$

The transformer is described as 'step-up' or 'step-down' according to whether A or B, respectively, is used as the primary (where $N_1 < N_2$).

Assuming perfect efficiency, Power supplied = Power delivered, or

$$E_1 I_1 = E_2 I_2 \quad ....(40.5)$$

so if the voltage is stepped up, the current is reduced.

The current taken from the secondary affects the current required in the primary. Both primary and secondary currents cause a flux through *both* coils, but these act in opposite directions: the flux of each coil tends to *reduce* its own current (by Lenz's law), but to *induce*, or increase, the current in the other circuit. Thus an increase in the secondary current causes an increase in the primary current.

41. Chemical and Thermoelectric Effects

ELECTROLYSIS

An *electrolyte* is a liquid that conducts electricity, undergoing some chemical change in the process. *Electrolysis* is the process of the conduction of electricity in an electrolyte. Two or more electrodes in an electrolyte constitute a *voltameter* or *coulometer*.

ELECTRICITY

The ionic theory. When salts, etc. that form good electrolytes are dissolved, almost all the molecules dissociate into oppositely charged particles, called *ions*. For example, sodium chloride dissociates into the ions Na^+ and Cl^-, copper sulphate into Cu^{++} and SO_4^{--}. The evidence is that this occurs on solution, and not *after* a p.d. is applied across electrodes placed in the solution.

When a p.d. *is* applied, these ions drift towards the electrodes, and a current flows in the solution. The liberation of substances at the electrodes is due to the giving up of the charges by these ions on arrival, the ions thereby reassuming the chemical properties of the uncharged substances.

By this theory Faraday's laws are explained. Since each ion carried a fixed mass and a fixed charge, the mass of substance liberated must be proportional to the quantity of electricity passed (first law).

Each ion carries n electronic charges, where n is the valency of the ion. The passage of 1 mole of ions (liberating the molar mass M of an element) thus carries across n mole of electronic charges. A mass M/n of any element liberated thus carries across 1 mole of electronic charges, which is a constant quantity (hence second law).

If $\quad N_A$ = Avogadro constant = 6.022×10^{23} per mole
and $\quad e$ = Electronic charge = 1.602×10^{-19} coulomb,
then $\quad F$ = Charge on 1 mole of electrons
$\quad\quad\quad$ = Faraday constant = 9.650×10^4 coulomb per mole,

since
$$F = N_A \cdot e \quad\quad\quad\quad\quad\quad\quad (41.1)$$

This equation provides a means of obtaining a value for the electronic charge e, if F is measured in electrolysis, and N_A is known from other sources.

Laws of electrolysis

FARADAY'S LAWS OF ELECTROLYSIS. (1) The mass m of any substance liberated in electrolysis is proportional to the current I and to the time t, i.e. to the quantity of electricity It passed.

Thus $m \propto It$, $\quad$ or $\quad m = zIt \quad\quad\quad\quad\quad\quad (41.2)$

where z is the electrochemical equivalent of the substance.

(2) When the same current flows for the same time through a series of different electrolytes the masses of substances liberated are in the ratio of their respective values of $\dfrac{\text{Molar mass}}{\text{Valency}}$

198

Thus
$$\frac{m_1}{m_2} = \frac{M_1/n_1}{M_2/n_2} = \frac{z_1}{z_2} \quad \text{...............(41.3)}$$

where the suffixes 1 and 2 refer to two substances liberated by the same quantity of electricity, M_1 and M_2 being their respective molar masses in kg mol^{-1}, and n_1, n_2 their respective valencies.

The ELECTROCHEMICAL EQUIVALENT (z) of a substance is the mass liberated in electrolysis per unit quantity of electricity passing. *Unit:* kg C^{-1}.

The FARADAY CONSTANT (F) is the quantity of electricity required to liberate the $\dfrac{\text{Molar mass}}{\text{Valency}}$ of any substance in electrolysis. It is also the charge on 1 mole of electrons.

Faraday constant $= 9.650 \times 10^4$ C mol^{-1}.

Energy considerations in electrolysis. The power equation is
$$VI = V'I + I^2R \quad \text{...............(41.4)}$$

where VI is the power supplied, $V'I$ the power used in maintaining the chemical actions at the electrodes, and I^2R the power used in heating the electrolyte, of resistance R.

Thus V' is a 'back e.m.f.' (p. 157), or potential drop, due to the conversion of electrical energy into chemical energy. It is the number of joule required to maintain the chemical actions per coulomb of electricity passed.

Dividing Eqn. (41.4) by I we obtain
$$V - V' = IR \quad \text{...............(41.5)}$$

No current should therefore flow until the p.d. V applied across the plates exceeds the back e.m.f. V', but thereafter the graph of I against V should be linear (Fig. 191).

A rough value for V' can be calculated as follows. In the case of the water voltameter, the chemical energy required is that to decompose water. By experiment, the combination of the Molar mass/ Valency of hydrogen and oxygen into water liberates 1.40×10^5 joule of heat. Therefore, to decompose these same quantities in electrolysis

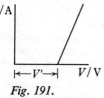

Fig. 191.

requires 1.40×10^5 joule of electrical energy. But this is accompanied by the passage of 9.65×10^4 coulomb (the Faraday constant).

$$\therefore \quad \text{P.d. } V' \text{ required} = \frac{1.40 \times 10^5 \text{ joule}}{9.65 \times 10^4 \text{ coulomb}} = 1.48 \text{ volt.}$$

V' is found, by experiment, to be actually 1.7 volt.

CELLS

When two substances are brought into close contact, for equilibrium to exist electrons must be transferred from one to the other until a certain p.d. has built up across the boundary. A similar p.d., called an *electrode potential*, is formed across the boundary between a metal electrode and an electrolyte.

If two *similar* electrodes are placed in an electrolyte, making a voltameter, the two electrode potentials are equal and opposite, and there is no net p.d. across the plates. If the electrodes are of *different* metals, the electrode potentials are different, and a net e.m.f. exists across the plates: this arrangement forms a voltaic cell. A conductor connected externally across the plates allows a current to flow down this p.d. This movement of electrons disturbs the equilibrium at the electrode-electrolyte boundaries, and ions enter or leave the electrolyte to restore it, causing a current through the liquid, and the associated chemical effects.

Energy considerations in cells. When a Daniell cell supplies a current, zinc goes into solution at the negative plate, and copper is deposited at the positive plate. By calorimeter experiments, it can be shown that the heat of formation of the Molar mass/Valency of $ZnSO_4$ from zinc is 7.90×10^4 joule, and of $CuSO_4$ from copper is *minus* 2.59×10^4 joule.

Therefore the passage of 9.65×10^4 coulomb through a Daniell cell should yield $(7.90 \ plus \ 2.59) \times 10^4$ joule of electrical energy. This is the source of energy of the cell, and the calculated e.m.f. should be

$$E = \frac{1.05 \times 10^5 \text{ joule}}{9.65 \times 10^4 \text{ coulomb}} = 1.08 \text{ volt.}$$

THERMOELECTRICITY

SEEBECK EFFECT. When two metals are joined to form a complete circuit, and one junction is maintained at a higher temperature than the other, an e.m.f. is set up in the circuit, and a current flows.

The magnitude of the e.m.f. obtained depends upon what metals form the junctions, and upon the temperature of each junction. A potentiometer can be used (see p. 167) to obtain curves such as those shown (Fig. 192). The curves A and B refer to cold-junction temperatures of 0°C and 100°C, respectively,

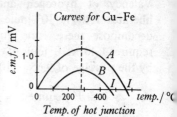

Curves for Cu–Fe

Temp. of hot junction

Fig. 192. Thermoelectric curves.

200

41 . CHEMICAL AND THERMOELECTRIC EFFECTS

for various temperatures of the hot junction. Each curve attains a maximum at the *neutral temperature* —285°C for copper-iron—and the e.m.f. reverses direction at points *I*, called *inversion temperatures*. The effect is used in the thermoelectric thermometer (p. 130).

PELTIER EFFECT. When a current is passed across the junction of two metals, the junction either cools or heats up, depending on which way the current flows.

This is the complementary effect to the Seebeck effect and indicates that an e.m.f. exists at every junction between two metals. A current flowing in the direction in which the e.m.f. is acting (from L. to R., Fig. 193*a*) experiences a 'step up' in potential, and acquires energy. This energy can come only from the junction itself, which consequently cools. A current flowing in the opposite direction to the e.m.f. (from R. to L., Fig. 193*b*) suffers a 'step down' in potential, and gives out energy. This appears as heat, and the junction heats up.

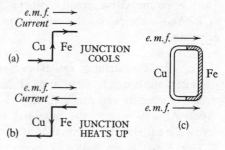

Fig. 193. Peltier effect.

The existence of these junction potentials accounts for the Seebeck effect. If the two junctions are at the same temperature, two equal Peltier e.m.f.s act in opposite directions, and no current flows (Fig. 193*c*). When one junction is heated, its e.m.f. increases, since the magnitude of the e.m.f. depends upon temperature. A net e.m.f. then acts in the circuit. Other e.m.f.s are, in fact, also set up, due to the temperature gradients along the two wires.

Simple demonstration of Seebeck and Peltier effects. With the switch to *A* (Fig. 194), a current flows clockwise in the lower circuit, cooling junction *C* and heating junction *D*, by the Peltier effect. The switch is now thrown to *B*, and, since *D* is hotter than *C*, a thermo-electric current flows clockwise in the upper circuit, by the Seebeck effect, until the temperatures have again become equal.

Fig. 194.

42. Alternating Current

Production of alternating current by rotating coil. A coil (N turn, area A) rotates at an angular velocity ω in a uniform field of flux density B (Fig. 195). At any instant its plane makes an angle ωt with the field, where t is the time in second.

Instantaneous flux linkage Φ of the coil $= BAN \sin \omega t$. The instantaneous e.m.f. E generated is $d\Phi/dt$, and is obtained by differentiating w.r.t. the time t,

$$E = BAN\omega \cos \omega t = E_0 \cos \omega t \quad \ldots\ldots\ldots\ldots(42.1)$$

Thus a cosine curve results, where E_0, or $BAN\omega$, is the peak value of the e.m.f. generated.

A.c. circuit containing resistance only. An alternating e.m.f. E is applied in a circuit containing resistance R only (Fig. 196a). At any instant the potential rise across E is equal to the potential drop across R. Thus $E = V$ at any instant. The current I through R is proportional to V (Ohm's law), so I is alternating in phase with V.

Thus in an a.c. circuit the current is *in phase* with the p.d. across the resistance.

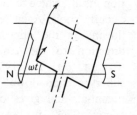

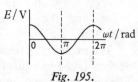

Fig. 195.

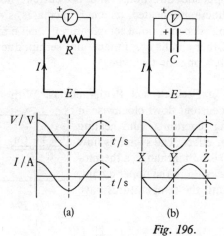

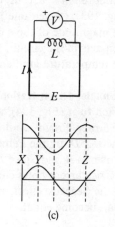

Fig. 196.

A.c. circuit containing capacitance only. An alternating e.m.f. E is applied in a circuit containing capacitance C only (Fig. 196b). As in the previous case, $E = V$, the p.d. across the capacitance, at every instant.

The charge Q on the plates is at every instant proportional to V. At point X in the cycle V is maximum, and the charge Q is maximum, with the $+$ charge on the L.H. plate as shown. Between X and Y in the cycle the $+$ charge flows in the negative direction to discharge the capacitor and then recharge it in the opposite sense. During this period the current I is therefore negative. At Y the current reverses, again discharging and recharging the capacitor, and the cycle is completed at Z.

The peak current thus occurs *in advance of* the nearest peak in the p.d. curve, and the current *leads* the p.d. across the capacitance by a quarter of a cycle.

A.c. circuit containing inductance only. An alternating e.m.f. E is applied in a circuit containing inductance L only (Fig. 196c). The current in the circuit is changing, and so is the flux Φ linking the circuit. A back e.m.f. V, equal to $d\Phi/dt$, is thus induced at every instant across the inductance L. This back e.m.f. is alternating, since at every instant $E = V$.

When I is increasing at the maximum rate, at point X in the cycle, the back e.m.f. V tending to oppose the change is a maximum. At point Y, the current is maximum and is momentarily steady, and V is zero, since the flux is not changing. As the current decreases, V changes sign. The cycle is completed at Z.

The peak current thus ocurs *after* the nearest peak in the p.d. curve, and the current *lags behind* the p.d. across the inductance by a quarter of a cycle.

R.m.s. current and e.m.f.

The ROOT MEAN SQUARE (or 'VIRTUAL') current is that steady current which gives the same heating effect as the alternating current.

An alternating current $I = I_0 \cos \omega t$ is passed through a resistance R. Heat developed in time $\delta t = I^2 R\, \delta t = I_0{}^2 R \cos^2 \omega t \cdot \delta t$. Heat developed in one complete cycle which takes time t

$$= I_0{}^2 R \int_0^t \cos^2 \omega t \cdot dt = I_0{}^2 R \int_0^t \tfrac{1}{2}(\cos^2 \omega t + \sin^2 \omega t)dt,$$

since the integrals of $\cos^2$ and $\sin^2$ over a complete cycle are equal.

$\therefore$ Heat developed $= \dfrac{I_0{}^2}{2}Rt$, where I_0 is the *peak current*.

But, by definition, Heat developed $= \overline{I^2}Rt$, where $\overline{I^2}$ is the *mean square current*.

$$\therefore \qquad \text{R.m.s. current} = \frac{\text{Peak current}}{\sqrt{2}} \qquad \ldots\ldots\ldots\ldots(42.2)$$

Thus (Fig. 197) the mean square current is *half* the peak value I_0^2, and the root mean square current is $\dfrac{1}{\sqrt{2}}$ the peak value I_0. The area beneath the I^2 curve is proportional to the heating effect, and is equal to the area beneath the mean square current line.

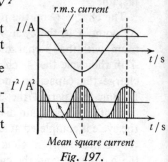

Fig. 197.

Similarly,

the ROOT MEAN SQUARE (or 'VIRTUAL') E.M.F. is that steady e.m.f. which gives the same heating effect as the alternating e.m.f.

By identical reasoning we obtain

$$\text{R.m.s. e.m.f.} = \frac{\text{Peak e.m.f.}}{\sqrt{2}} \qquad \ldots\ldots\ldots\ldots(42.3)$$

Power in the a.c. circuits

The power developed across a capacitance at any instant $= VI$. However, we have seen that in an a.c. circuit I leads V by a quarter of a cycle. Thus if

$$V = V_0 \sin \omega t, \qquad \text{then} \qquad I = I_0 \sin(\omega t + \frac{\pi}{2}) = I_0 \cos \omega t.$$

Energy developed in one cycle which takes time t

$$= \int_0^t VI dt = V_0 I_0 \int_0^t \sin \omega t \cos \omega t \,.\, dt = V_0 I_0 \int_0^t \tfrac{1}{2} \sin 2\omega t \,.\, dt.$$

But the integral of the sine function over a complete cycle is zero, therefore this integral is zero.

Therefore in an a.c. circuit the power developed across a capacitance is zero. Similarly, the power developed across an inductance is zero. It is only the resistive component that develops power, and this is dissipated in the form of heat.

Factors influencing magnitude of current. In an a.c. circuit the magnitude of the current depends upon the following constants, in addition, of course, to the magnitude of the applied alternating e.m.f.

(1) As the resistance R increases, the current decreases.

(2) The larger the capacitance C, the greater the charge flowing to and from the plates during each cycle, and the greater the current. The higher the angular frequency ω in a capacitive circuit, the more times per second the charge flows back and forth, and the greater the current.

(3) The larger the inductance L, the smaller the current required to produce the necessary flux linkage and back e.m.f. to balance the forward e.m.f. Thus a self-inductance acts as a 'choke' to the current. The higher the angular frequency ω in an inductive circuit, the greater the rate of change of current, and the smaller the current required to produce the back e.m.f.

Part VI
Modern Physics

43. Electrons and Positive Ions

THE ELECTRON

The electron as a particle

The term *electron* originally denoted simply the basic unit of charge. It soon came to mean, however, a definite particle, having both charge and mass, and capable of independent existence outside of other matter. It is in this sense that we seek evidence of its existence, and information regarding its charge and mass and other properties. The electron is normally bound within the atom, but can be liberated by various means and observed in its free state. Depending upon the means of liberation, free electrons are referred to variously as:

> Thermionic electrons,
> Cathode rays,
> Photoelectrons, or
> Beta particles.

Measurements of the charge e and specific charge e/m of these various particles, produced in different ways, show them to be all identical in nature.

The thermionic effect

The easiest way to produce free electrons is to heat a metal filament by passing a current through it. The filament should be in a vacuum, since, although a metal heated in the atmosphere would emit electrons, the effect would be considerably confused by the surrounding gas. Other materials, such as various oxides, also exhibit thermionic emission.

THERMIONIC EMISSION is the release of electrons from a body by reason of its temperature.

In a conductor, some of the electrons are not bound to particular atoms but behave rather like an electron 'gas' within the confines of the conductor. These are the electrons which give electrical conducting properties to a metal. Heating the conductor increases the thermal energies of these electrons, and some of them are then able to break free from the surface. The process is akin to evaporation and, as in the case of evaporation, the rate of emission of electrons increases with temperature.

43 . ELECTRONS AND POSITIVE IONS

Qualitative evidence for thermionic electrons

To investigate the thermionic effect a highly evacuated glass vessel *D* containing two metal electrodes—an anode plate *A* and a tungsten filament cathode *C*—is used (Fig. 198). The filament is heated by a circuit containing a 4- or 6-V battery *B* and a rheostat *R* to vary the filament temperature. A p.d. V_a of up to about 400 V is applied across *AC* from a variable d.c. source, and the resulting current *I* through the 'valve' is measured by a milliammeter.

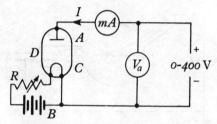

Fig. 198. Diode with directly-heated filament.

Effect of filament temperature. It is found that no current *I* flows unless *C* is heated, and the higher the temperature the higher the current. This is explained in terms of the cathode *C* emitting negative electrons, which are drawn across to the positive anode *A*, producing a negative current from *C* to *A* through the valve. This is equivalent to a positive current *I* in the reverse direction.

Effect of reversing the potential. If the anode is now made negative instead of positive, no current flows, even when the filament is heated. For a current to flow in this case, electrons would have to be emitted from *A*, and drawn across to *C*. But since *A* is not heated, no electrons are available there. Hence the term 'valve' for this device, which allows current to flow in one direction only.

The 'Maltese cross' experiment. If a much larger p.d. (from 3 to 5 kV) is applied across a suitable tube, the electrons are accelerated to a very high velocity. On striking the wall of the tube beyond the anode they are seen to cause the glass to fluoresce. An obstacle, conventionally shaped like a Maltese cross, placed in the path of the electrons, casts a geometrical shadow. This shows that the fast electrons travel from the cathode in straight lines.

The effect can also be produced using a cold cathode, if the p.d. across the tube is sufficiently high, i.e. several thousand volt. It was in cold cathode discharge tubes that fast electrons were first produced, and were then named 'cathode rays'.

209

MODERN PHYSICS

Deflection by a magnetic field. If the rays are indeed negatively charged particles, they are equivalent to an electric current. They should therefore be deflected in a magnetic field according to Fleming's L.H. motor rule (p. 169), remembering, of course, that the electron flow is in the opposite direction to conventional current. The electrons are found, in fact, to be deflected according to this rule.

Perrin's tube. One form of this tube is shown in Fig. 199. Thermionic electrons from the heated filament C are accelerated to a

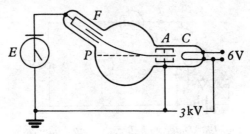

Fig. 199. Perrin's tube.

high velocity by a potential of about 3 kV on the anode A. They pass through an aperture in A and fall on the wall of the tube at P, causing fluorescence. A magnetic field perpendicular to the plane of the diagram is provided by external coils, and when this is switched on and adjusted to the right value, the electron beam is deflected into a Faraday cylinder F connected to an electroscope E. The electroscope is found to become negatively charged when the beam is deflected in this way. The experiment provides convincing evidence that the beam consists of negatively charged particles.

Diode characteristics

The cathode (Fig. 198) may be either a directly heated tungsten filament, oxide-coated to increase the thermionic emission of electrons; or alternatively, an oxide-coated cylinder heated indirectly by an insulated filament running through it.

Electrons are 'evaporated' from the cathode with velocities of emission which vary with individual electrons. A negative space charge exists in the vicinity of the cathode, consisting of the electrons just emitted. This space charge repels those electrons possessing only small thermal velocities back to the cathode. Some of the electrons, however, with higher initial velocities, are able to break through the space charge, and if the anode is made positive with respect to the cathode these electrons will be attracted there, producing an anode

210

current I. As the anode voltage V_a is increased, the adverse electric field due to the space charge is reduced. Consequently, more electrons are able to break through and the anode current rises (O to B in Fig. 200).

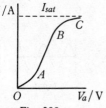

Fig. 200.
Curve for diode.

Eventually, when V_a is sufficiently large to overcome entirely the space charge effect, *all* the electrons emitted from the cathode reach the anode (point C). Since no more electrons are available no further increase in current can occur. The value of the 'saturation current' obtainable therefore depends on the total emission, which is governed only by the filament temperature.

The curve shown applies to a hard valve. A hard valve has the highest possible vacuum; a soft valve contains an easily ionized gas, such as argon, at reduced pressure. The curves for soft valves are similar in general shape, but beyond point A an additional effect occurs. As the electrons move towards the anode, collisions take place with the gas molecules present. As V_a is increased, the electron velocities increase, and reach values at which the collisions cause ionization of the gas. When the gas is ionized, the positive ions produced, being more massive, move much more slowly to the cathode than the electrons do to the anode. They are therefore in the vicinity of the negative electron space charge long enough to neutralize it and so permit much larger anode currents. Thus beyond point A, at which this effect occurs, the current in a soft valve is much higher than that in a similar hard valve. Soft valves are used for heavy duty rectifiers, etc.

Diode as rectifier. We have seen that a valve passes current in one direction only—an electron flow from cathode to anode. No current can flow in the reverse direction if an opposite p.d. is applied, since no corresponding current-bearing electrons are emitted thermally from the anode.

Fig. 201a shows the principle of half-wave rectification. An alternating e.m.f. is applied through a transformer to the circuit containing the diode. A half-wave rectified current is produced (Fig. 201b), the valve acting as an insulator during the half-cycle in which the cathode is positive with respect to the anode. The output can be smoothed, as shown, by a condenser C.

A 'double diode' circuit (Fig. 201c) can effect full-wave rectification. Either anode A or anode B is, at a given instant, positive with

211

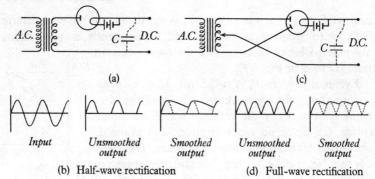

(a)   (c)

| Input | Unsmoothed output | Smoothed output | Unsmoothed output | Smoothed output |

(b) Half-wave rectification   (d) Full-wave rectification

Fig. 201. Diode as rectifier.

respect to the cathode, since the cathode is tapped at the centre of the secondary coil. The electrons will therefore flow from the cathode to A and B alternately, and the output current will be as shown (Fig. 201d). A condenser C achieves a smoother output.

Triode characteristics

In the triode, a third electrode in the form of a grid is introduced between cathode and anode. By varying the voltage on the grid the electron current from cathode to anode is very sensitively controlled. If the grid is made more negative, the space charge effect is increased and the anode current reduced. Conversely, a more positive grid increases the anode current. The anode current I_a depends upon both the anode voltage V_a and the grid voltage V_g (both taken with respect to the cathode), but since the grid is closer to the cathode than is the anode, it is more effective in controlling the current.

A series of curves may be obtained using the circuit in Fig. 202a. If the grid is connected directly to the cathode, then $V_g = 0$, and the curve of I_a against V_a is similar to that of the diode (Fig. 200). It is, however, more useful to plot curves of I_a against V_g for various constant values of V_a (Fig. 203), the voltage V_g being varied from negative to positive values using tappings from a grid-bias battery. It will be seen that each curve has a linear portion, these portions being parallel to one another. From these curves we may define certain valve parameters, which will be constant if the valve is operated within this linear region.

(1) The AMPLIFICATION FACTOR (μ)

$$= \frac{\text{Change in anode voltage to produce given change in anode current}}{\text{Change on grid voltage to produce same change in anode current}}.$$

We have seen that the grid controls the current more sensitively than does the anode, and this makes possible the use of the triode as an amplifier. For example, in Fig. 203, an increase in I_a of 1 mA can be achieved either by increasing V_a from 100 to 120 V, keeping

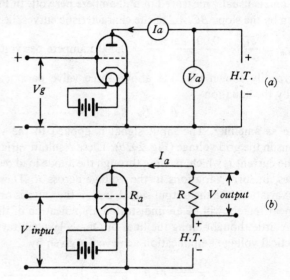

Fig. 202. (a) Triode with indirectly-heated filament. Circuit for triode characteristics. (b) Triode as amplifier.

V_g constant at -2 V (point C to B), or by changing V_g from -4 to -2 V, keeping V_a constant at 120 V (point A to B). Thus,

$$\mu = \frac{\delta V_a}{\delta V_g} = \frac{20 \text{ volt}}{2 \text{ volt}} = 10.$$

(2) The ANODE SLOPE RESISTANCE (R_a)

$$= \frac{\text{Change in anode voltage}}{\text{Corresponding change in anode current}}.$$

A valve does not obey Ohm's law (see Fig. 200), but by considering *changes* in V_a and I_a along the linear part of the curves a 'differential' or 'a.c.' resistance of the valve can be defined as above. Thus, in Fig. 203, a change in V_a of 20 V, at a constant V_g of -2 V (point C to B), corresponds to a change in I_a of 1 mA, giving

$$R_a = \frac{\delta V_a}{\delta I_a} = \frac{20 \text{ volt}}{0.001 \text{ ampere}} = 20\ 000 \text{ ohm}.$$

213

(3) The MUTUAL CONDUCTANCE (g_m)

$$= \frac{\text{Change in anode current}}{\text{Change in grid voltage producing it}}.$$

This parameter indicates the influence of the grid on the anode current, and is usually measured in milliampere per volt. In Fig. 203, it is given by the slope BC/AC of the characteristic curves, i.e.

$$g_m = \frac{\delta I_a}{\delta V_g} = \frac{0.001 \text{ ampere}}{2 \text{ volt}} = 0.000\ 5 \text{ ampere per volt.}$$

It is readily shown that the above three valve parameters are related by the equation

$$\mu = R_a \cdot g_m \quad \dots\dots\dots\dots\dots\dots\dots(43.1)$$

Triode as amplifier. The input signal is applied to the valve as variations in the grid voltage (Fig. 202b). These result in variations in the anode current I_a which, passing through the anode load resistance R, causes, in turn, variations in the voltage across R. These latter voltage variations are the output signal, and, if the valve is operating in its linear region, will be an undistorted amplification of the input signal. A little thought along the lines of Ohm's law will reveal that the practical voltage amplification achieved is given by

$$\frac{\delta V_{output}}{\delta V_{input}} = \mu \left(\frac{R}{R + R_a} \right) \quad \dots\dots\dots\dots\dots(43.2)$$

where μ is the amplification factor, R_a the anode slope resistance, and R the anode load resistance. The larger R is in relation to R_a, the

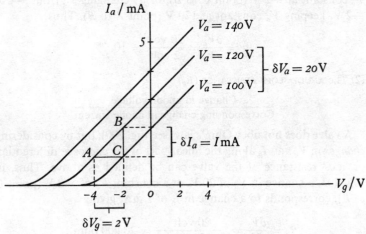

Fig. 203. Triode characteristics.

greater the amplification achieved. Too large a value of R, however, would reduce I_a excessively. If R were made 80 000 ohm, in circuit with the triode characterized in Fig. 203, the amplification would be

$$\mu\left(\frac{R}{R + R_a}\right) = 10\left(\frac{80\ 000}{100\ 000}\right) = 8.$$

The cathode ray tube

Electrons are emitted thermally from an oxide-coated cathode C, heated by an adjacent filament, in an evacuated glass envelope (Fig. 204).

The Wehnelt cylinder W is at a negative potential (say -50 V) with respect to the cathode, and repels the electrons—which would otherwise diverge—into a concentrated beam which passes through

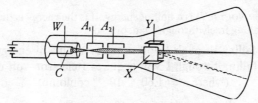

Fig. 204. Cathode ray tube.

the first aperture of the anodes. The brilliance, or intensity of the beam, depends on the potential of W; if this is made sufficiently negative the beam can be cut off altogether.

The electrons are accelerated by the high potentials of the anodes A_1 (say $+800$ V) and A_2 (say 2 kV). The anodes also form an 'electron lens' which focuses the beam on to the screen; focusing can be adjusted by varying the potential of A_1.

The beam can be deflected electromagnetically by external coils, or electrostatically by internal plates, as shown. By varying the p.d.s across these X and Y plates the beam can be made to strike the screen at any point, causing fluorescence and a spot of light at that point.

Voltage required to accelerate electrons

In the cathode ray tube the electrons attain velocities of the order of one-tenth the speed of light. The p.d. required to accelerate electrons in a vacuum to this speed can be calculated as follows.

An electron of charge e falling through a p.d. V loses potential energy eV (Eqn. 34.5). Kinetic energy it gains is $\frac{1}{2}mv^2$, where m is its mass and v the velocity it attains.

Thus $$eV = \tfrac{1}{2}mv^2 \dots\dots\dots\dots\dots\dots\dots(43.3)$$

The assumption is that e/m for an electron remains constant. As the speed of light is approached, this assumption breaks down, but the error is only about 1% in the present case.

$$e = 1.60 \times 10^{-19}\,\text{C}$$
$$m = 9.11 \times 10^{-31}\,\text{kg}$$
$$v = 3.00 \times 10^7\,\text{m s}^{-1}$$
$$\therefore \quad V = 2560\,\text{volt.}$$

The electron-volt. This unit is to fall into disuse in the SI system, but may still be met in atomic physics. It is seen to be related to the joule as follows:

The ELECTRON-VOLT (eV) is a unit of energy. It is the energy acquired by an electron in falling freely through a p.d. of 1 volt.

1 coulomb falling through 1 volt acquires 1 joule of energy
e coulomb falling through 1 volt acquires 1 electron-volt of energy
(where $e = 1.60 \times 10^{-19}$ coulomb)

Thus 1 electron-volt $= 1.60 \times 10^{-19}$ joule.

Electrons in an electric field

An electron placed between two parallel plates across which a p.d. is applied will be deflected towards the positive plate. If the electric field between the plates is E and the electronic charge is e, the force F on the electron is given by Eqn. (38.4)

$$F = Ee \dots\dots\dots\dots\dots\dots\dots(43.4)$$

Electric field between parallel plates. The uniform electric field E in the above expression is given by Eqn. (38.6)

$$E = \frac{\text{P.d. across plates}}{\text{Distance between plates}} = \frac{V}{d} \dots\dots\dots(43.5)$$

Motion of electrons in a uniform magnetic field

A stationary electron is unaffected by a magnetic field, but a moving electron constitutes an electric current, and a force is therefore exerted on it by a magnetic field, in accordance with Fleming's L.H. motor rule (p. 169). The force will be at right-angles to the direction of motion of the electron.

It will be recalled (Eqn. 36.3) that the force F on a wire of length l

216

carrying a current I in an external magnetic field of flux density B, is given by

$$F = BIl,$$

where the current, field and force vectors are mutually at right-angles.

Now consider an electron, of charge e, travelling in a straight line at a velocity v. In one second, therefore, a charge e has passed every point along a distance v in metre, and this is equivalent to a negative current e in coulomb per second, or ampere, in a wire of length v in metre. In the above equation, it follows that we can substitute, numerically, e for I and v for l, giving

$$F = Bev \quad\dots\dots\dots\dots\dots\dots\dots(43.6)$$

for the force F on a charge e travelling at a velocity v in an external magnetic field of flux density B.

In a transverse magnetic field, the electron will thus experience a constant force always perpendicular to its direction of motion. This is exactly the condition required for circular motion (p. 7), the force Bev providing the required centripetal force $\dfrac{mv^2}{r}$ directed towards the centre of a circle of radius r, where m is the mass of the electron in kilogramme. While the electron remains within a uniform field, therefore, its path will be circular, given by

$$Bev = \frac{mv^2}{r}\dots\dots\dots\dots\dots\dots\dots(43.7)$$

If an appreciable part of the circle is completed by the electron, it may be possible to measure this radius r directly. If, on the other hand, only a small segment of the circle is transversed within the magnetic field, the radius may be deduced from measurements of the length x and linear deflection h (Fig. 205). The radius is given geometrically (see footnote, p. 30) by

$$r = \frac{x^2 + h^2}{2h} \quad\dots\dots\dots\dots\dots\dots(43.8)$$

Magnetic field at centre of Helmholtz coils. A pair of similar coils is mounted coaxially at a distance apart equal to their common radius (Fig. 206). This is called the Helmholtz arrangement, and its importance is that it produces a uniform and calculable magnetic field over a considerable region around the centre of the system. It is therefore often used for the present purpose.

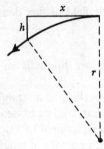

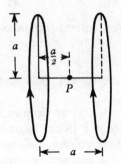

Fig. 205.   Fig. 206. Helmholtz coils.

The field at P due to each coil is given by Eqn. (36.9),

$$\tfrac{1}{2}B = \frac{\mu_0 N I a^2}{2(a^2 + x^2)^{3/2}}$$

where $x = \tfrac{1}{2}a$. Therefore total field due to both coils is

$$B = \frac{8\mu_0 N I}{5\sqrt{5} \cdot a} \quad\ldots\ldots\ldots\ldots\ldots\ldots\ldots(43.9)$$

Measurement of e/m and v of electrons

The above theory suggests possibilities for measuring the specific charge e/m and velocity v of electrons, though it does not provide methods for e and m separately. Two school methods are here described. The theory can be readily adapted to the other methods available: viz., those using the 'magic eye', and magnetron principles.

Method I, using the fine beam tube. A stream of electrons from a thermionic cathode is accelerated to the anode which is at the voltage V and passes through a hole in the anode (Fig. 207b), having attained a velocity v given by $eV = \tfrac{1}{2}mv^2$ (Eqn. 43.3).

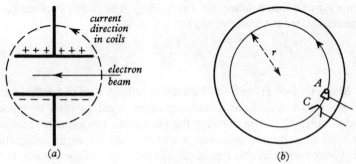

Fig. 207. (a) Crossed electric and magnetic fields. (b) Fine beam tube.

43 . ELECTRONS AND POSITIVE IONS

Being no longer subjected to the electric field between cathode and anode the electrons continue at this velocity, and are subsequently deflected by a uniform magnetic field of flux density B, applied by external Helmholtz coils, into a circular path of radius r given by

$$Bev = \frac{mv^2}{r} \text{ (Eqn. 43.7).}$$

The electrons are contained in a concentrated beam by 'gas focusing', and their circular path is made visible by ionization of the residual gas in the tube. The electrons traverse an almost complete circle, and their radius r can thus be measured directly. Combining the above equations, we obtain values for e/m and v in terms of the measurable quantities V, B, and r,

$$\left(\frac{e}{m}\right) = \frac{2V}{B^2r^2} \quad \text{and} \quad v = \frac{2V}{Br}.$$

Method II, using 'crossed' electric and magnetic fields. This is a simplified version of one of J. J. Thomson's original experiments.

(1) In this case, the velocity v is obtained directly, without reference to the accelerating voltage V. An electric field E and a magnetic field B are both applied to the electron stream beyond the anode, in such a way that their effects act in opposition and exactly cancel, resulting in zero deflection of the beam (Fig. 207a).

If the electric field is vertical, with the positive plate uppermost, an electrostatic force $F = Ee$ (Eqn. 43.4) acts upwards on the electrons. To counteract this, the magnetic field from the Helmholtz coils must be applied *perpendicular* to the electric field as shown. By Fleming's L.H. motor rule, this gives an electromagnetic force $F = Bev$ (Eqn. 43.6) downwards on the electrons.

These fields are adjusted until the beam is undeflected. Assuming that they are both *uniform*, and applied over the same region, then

$$Ee = Bev \quad \dots\dots\dots\dots\dots(43.10)$$

giving for the velocity of the electrons $v = E/B$.

(2) A deflection is then obtained by applying a magnetic field B' alone. The linear coordinates x and h of the deflected beam are noted (Fig. 205). Applying $B'ev = \frac{mv^2}{r}$ (Eqn. 43.7) we obtain

$$\left(\frac{e}{m}\right) = \frac{v}{B'r},$$

219

where r is deduced from $r = \dfrac{x^2 + h^2}{2h}$ (Eqn. 43.8) and v is already known from the first part of the experiment.

The electron as a universal constituent of matter

Before the discovery of the electron, the smallest known particle was the hydrogen ion, later called the proton. To determine the relationship between these two particles, it was necessary to be able to measure independently the mass and charge of both.

The mass m' of the hydrogen atom. The nineteenth century produced not only convincing evidence of the existence of atoms, but also methods of estimating their actual sizes and masses. The mass of the hydrogen atom was estimated at $m' = 1.6 \times 10^{-27}$ kg.

The charge e on an ion in solution. Faraday's laws of electrolysis gave the mass deposited or liberated as proportional to the charge carried, i.e., $M = zQ$ (Eqn. 41.2). This suggested that, since the mass M was particulate, the charge Q might be also. In which case, knowing the elementary ionic mass m' (e.g. of hydrogen, above), the ionic unit of charge e was easily calculated from $\dfrac{e}{m'} = \dfrac{Q}{M}$. The unit charge on an ion in solution was thus estimated at $e = 1.6 \times 10^{-19}$ coulomb.

The specific charge e/m' of the hydrogen ion. The above figures gave, for hydrogen,

$$\frac{e}{m'} = \frac{1.6 \times 10^{-19}}{1.6 \times 10^{-27}} = 10^8 \text{ C kg}^{-1}.$$

The specific charge e/m of the free electron. It was in the context of the above knowledge that the free electron was discovered in the 1890's. Its specific charge was measured at approximately

$$\frac{e}{m} = 2 \times 10^{11} \text{ C kg}^{-1}.$$

This was roughly 2000 times larger than the value for the hydrogen ion. It was necessary to discover whether this was due to the electron having a larger charge e or a smaller mass m than the hydrogen ion.

The charge e on a free electron. Using a method due to Townsend (now of historical interest only), involving the fact that water vapour will condense on charged gaseous particles to form visible drops, it was shown that the unit of charge for gaseous ions and free electrons, by whatever means produced—thermionically, by the photo-electric effect, etc.—was the same as that found for ions in solution, viz.

$$e = 1.6 \times 10^{-19} \text{ C}.$$

43 . ELECTRONS AND POSITIVE IONS

There was no longer any doubt of the existence of a universal basic unit of charge e.

The mass m of a free electron. It was now a simple matter to calculate the mass of the electron, which is now known accurately to be $\frac{1}{1838}$ that of the hydrogen atom. Also, from whatever source, only one kind of electron was found. The conclusion was that the electron is a building particle which is a universal constituent of all atoms.

Millikan's oil drop method for e

This famous experiment, performed in 1911, was the culmination of a series of attempts to obtain a more accurate value for e. It also illustrates well the particular nature of charge.

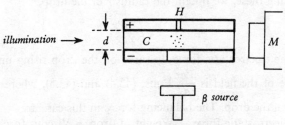

Fig. 208. Millikan's apparatus.

Very small drops from a spray are admitted through a small hole H into the draught-proof chamber C between two brass condenser plates, a known distance apart d (Fig. 208). The drops are observed through a microscope M which has been previously focused on a pin lowered through H. The drops are illuminated from a lamp-house, placed at a horizontal angle of about 165° to the microscope so as to provide maximum illumination by scattered light from the drop without the direct rays from the lamp entering the eye.

(1) A selected negatively-charged drop is allowed to fall freely between the plates, and its time of fall over a certain distance on the microscope graticule measured. Hence its velocity v_F of free fall is determined.

(2) A known p.d. V is now applied to the plates (upper plate positive) which is sufficient to cause the drop to rise at a velocity v_R. This velocity is determined in the same way as before.

Theory. A sphere of radius r moving at a velocity v in a medium of coefficient of viscosity η experiences an opposing viscous force F given by

$$F = 6\pi\eta rv \quad \dotfill (43.11)$$

This is known as Stokes' law, and is assumed to apply in the present case of a spherical oil drop moving through air. The drop therefore reaches a steady velocity such that the viscous force is equal and opposite to the other forces acting on the drop.

(1) Thus for the drop falling freely under gravity, Weight of drop = Viscous force. This gives

$$W = 6\pi\eta r v_F \dots\dots\dots\dots\dots(43.12)$$

If ρ is the density of the oil, the weight of the drop (ignoring Archimedes' buoyancy of the air displaced) is

$$W = \tfrac{4}{3}\pi r^3 \rho \dots\dots\dots\dots\dots(43.13)$$

Combining these, we obtain the radius r of the drop,

$$r = \sqrt{\frac{9\eta v_F}{2\rho g}}.$$

(2) The electrostatic force upwards on the drop rising under the influence of the field is $\dfrac{Ve}{d}$, Eqns. (43.4) and (43.5), where e is the charge on the drop. The balancing forces in this case are

Electrostatic force − Weight of drop = Viscous force,

or $$\frac{Ve}{d} - W = 6\pi\eta r v_R \dots\dots\dots\dots(43.15)$$

Adding Eqns. (43.12) and (43.14) we eliminate W, and re-arranging we obtain the charge e on the drop,

$$e = \frac{6\pi\eta r d}{V}(v_F + v_R).$$

To investigate more fully the particulate nature of the charge, a radioactive source of beta rays is applied temporarily to the drop, with the field switched off, causing it to change its charge. The measurement of v_R is then repeated. This procedure is repeated several times, with different charges on the same drop, and the multiple nature of the charge established. The value of unit charge is found to be 1.60 $\times$ 10^{-19} C.

POSITIVE IONS

Electrical discharge through gases

In the thermionic valve, the current-bearing electrons are made available by thermal emission from the hot cathode. An electrical discharge can, however, be obtained through a gas using cold electrodes. In this case, the discharge depends for its supply of electrons upon the ionization of the gas by collision. A sufficiently high potential is needed to enable the electrons to be accelerated to sufficiently high velocities to cause ionization. A reduced pressure may also be required, to give the electrons sufficient space to accelerate before colliding.

The various stages of the appearance of the discharge as the pressure is reduced are described in textbooks. At pressures of about 10^{-3} mmHg there are so few gas molecules that the electrons can accelerate right down the tube without collision, and their velocities on impact with the end of the tube are about one-tenth the speed of light, when the p.d. applied is about 2500 volt (p. 216). At these high energies they cause the glass to fluoresce and are called 'cathode rays'. High speed electrons produced by thermionic emission from a hot cathode have the same properties as these cathode rays, and their values of e/m are found to be the same. The conclusion is that these are identical particles. In the discharge tube the nature of the particles is found to be independent of the material of the cathode and of the gas present.

Positive rays in the discharge tube

A further phenomenon is observable in the discharge tube. If the cathode is perforated, a stream of radiation is seen on the side remote from the anode. This can be deflected by a very powerful magnetic field, the direction of deflection showing that it consists of positively charged particles. The deflection being much less than that of cathode rays, it follows that the particles are much heavier.

This is explained in terms of the mechanism of ionization of the gas. When a gas molecule is ionized an electron is released, leaving the molecule positively charged. Thus ionization always results in the creation of 'ion pairs'—in this case a negative electron and a heavier positive ion consisting of a molecule minus an electron. The positive ions, being more massive, are far less mobile than the electrons. The electrons are thus primarily responsible for the passage of current when a p.d. is applied.

Measurement of e/m of positive rays

The basic principles of measurement of e/m for positive rays are exactly the same as for electrons. Since there are two unknowns, e/m and v, as before, both magnetic and electric fields must be applied in order to obtain two equations. Since we can now assume that the positive ions possess a unit electronic charge e or a small integral multiple of it, the interest centres primarily on the measurement of the mass m. Instruments for this purpose are called mass spectrographs.

In the case of positive rays, however, there is an additional complication in that the positive ions are produced at different points in the discharge tube, and therefore have been accelerated to different velocities on arrival at the cathode. This velocity variation would simply cause a spreading out of the beam when subjected to electric and magnetic fields. Instruments must be designed to overcome this difficulty.

The Bainbridge mass spectrograph

This is one type of instrument designed for positive ray analysis. The positive rays from the discharge tube are admitted in a narrow pencil through the slits S_1 and S_2 (Fig. 209). Crossed magnetic and electric fields B and E are applied to the beam as described on p. 219. This device selects ions with a single velocity v which satisfies Eqn. (43.10) $Ee = Bev$: only these ions will remain undeflected and able to pass through the slit S_3.

A beam of ions of single velocity v thus emerges into a second uniform magnetic field B', perpendicular to the plane of the diagram, which deflects them into a circular path given by Eqn. (43.7)

$$B'ev = \frac{m'v^2}{r}.$$

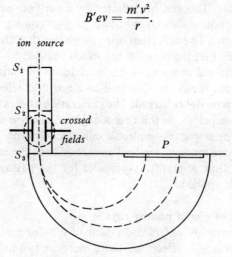

Fig. 209. *Bainbridge mass spectrograph.*

The radius r of the path taken will depend upon the mass m. Thus different sets of particles with different m values will produce different lines on the photographic plate P.

The proton

As is to be expected, the masses of positive rays depend upon the gas present in the discharge tube. However, many elements were found to consist of particles of more than one mass. These we call 'isotopes' of the element (see p. 249). For example, neon, with a relative atomic mass of 20.2, was found to consist of a mixture of two isotopes of masses almost exactly 20 and 22. Chlorine, with a relative atomic mass of 35.46, gave lines corresponding to masses 35 and 37. Indeed, all these isotopes were found to have relative atomic masses which were close to integers. This 'whole number rule' suggests that all atoms might be built up of a common particle or particles of mass close to unity.

The lightest positive particle found in the discharge tube was the hydrogen atom carrying a single positive charge—the hydrogen ion —of approximately unit mass (relative to carbon-12, taken as 12 exactly). By 1920, this particle was recognized as a fundamental building particle of all atoms, and was named the 'proton'.

44. X-rays

Discovery

In 1895 Röntgen discovered a new type of radiation in the vicinity of the discharge tube, which he called X-rays. Its nature was finally established in 1912 by crystal diffraction experiments (p. 227) to be high energy electromagnetic waves of wavelength between 10^{-7} and 10^{-11} m.

Production

X-rays are produced by the sudden deceleration of fast-moving electrons when striking a target material. The process is thus the converse of the photoelectric effect, in which electrons are released by electromagnetic radiation.

The early types of X-ray tube used a cold cathode, but the modern Coolidge tube derives the electrons by thermionic emission from a heated cathode, as in a valve. In this latter type, the *intensity* and *hardness* of the X-rays can be controlled independently. The intensity depends on the rate of emission of electrons, which is controlled by the cathode temperature, i.e. by the current through the cathode heater. The hardness (below) depends on the energy of the electrons on impact, and this is controlled by the voltage applied across the tube.

The beam from the oxide-coated cathode is focused by the Wehnelt cylinder W on to the target T (Fig. 210). The target is of tungsten or molybdenum, embedded in copper to conduct the heat away, and is water- or oil-cooled, since about 99% of the electron energy is convered directly into heat. The vacuum is as high as possible. The tube has its own rectifier, so an a.c. source can be used. The whole is encased, except for a small window, by lead, to absorb stray radiation.

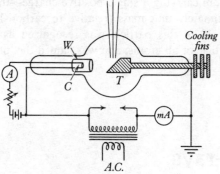

Fig. 210. X-ray tube.

Properties of X-rays

(1) They cause fluorescence.

(2) They affect photographic plates.

(3) They travel in straight lines.

(4) No deflection is observable in either magnetic or electric fields. This is in contrast to the behaviour of a stream of electrons.

(5) They ionize gases by the liberation of electrons.

(6) They ionize body tissue, and the effect is cumulative.

(7) They can penetrate matter to a degree depending on the wavelength. X-rays of short wavelength are said to be 'hard', and are more penetrating than 'soft' X-rays of longer wavelength.

(8) They can be reflected, refracted, diffracted, and polarized.

(9) They liberate electrons by the photoelectric effect.

Diffraction by crystals

In 1912, von Laue confirmed what was already suspected, that X-rays were of the same nature as light, but of much shorter wavelength. To measure such waves would need a diffraction grating having a spacing of the same order of magnitude as the wavelength, and existing ruled gratings were too coarse. Von Laue conceived the idea of using crystals, with their regular structure of atoms, as three-dimensional gratings for X-rays, since atomic spacing was known to be of the order of 10^{-10} m. By passing 'white' (continuous spectrum) X-rays through a crystal of zinc sulphide, he obtained on a photographic plate a diffraction pattern consisting of a number of symmetrical spots.

Shortly afterwards, W. L. Bragg suggested a simple way of calculating these diffraction maxima, in terms of refractions from crystal planes, i.e. planes intersecting a large number of atoms. Each atom scatters the X-rays, acting as a centre of secondary wavelets. A plane containing many atoms will therefore reflect strongly at the usual angle of reflection (by Huygens' principle, p. 76). Reflections from successive parallel planes will, however, interfere, and a maximum will occur only when these reflected rays interfere constructively.

Consider parallel X-rays, of wavelength λ, incident at a glancing angle θ on the two successive crystal planes P_1 and P_2, a distance apart s (Fig. 211). Reflection from each plane takes place at the same angle θ. The path difference between the two rays is $AB + BC = 2s \sin \theta$. For reinforcement, this must equal $n\lambda$ (Eqn. 16.1), where n is an integer. Reflected beams are therefore observed at angles θ given by

$$n\lambda = 2s \sin \theta \qquad \ldots\ldots\ldots\ldots(44.1)$$

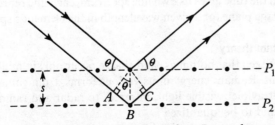

Fig. 211. Reflection of X-rays from Bragg planes.

which is known as the Bragg equation. If the spacing s between the given crystal planes is known, the wavelength λ can be found.

In later work, when accurate ruled gratings of known spacing were available, the X-ray wavelengths were obtainable directly from these. The Bragg equation could then be used to measure crystal spacings.

The Bragg X-ray spectrometer. In the Laue arrangement, the angles θ are pre-determined by the fixed orientation of the crystal. Each angle substituted in Eqn. (44.1) thus needs a certain definite wavelength for a spot to be formed. The experiment therefore depends on the continuous part of the X-ray spectrum (p. 230) to provide the various wavelengths necessary.

The Bragg spectrometer, on the other hand, can investigate any wavelength since θ can be adjusted to any desired value. The beam of X-rays from the tube is collimated by slits and reflects from the crystal planes to enter the detector (Fig. 212). The detector is an
ionization chamber, the principle of
which is discussed later (p. 240). The
crystal can be rotated to any angle,
the detector being always set to receive
the reflected beam. The Bragg angles θ
at which a response is obtained can thus
be measured with precision.

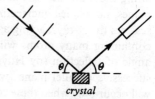

Fig. 212. Bragg X-ray spectrometer.

The powder method. In 1916 Debye and Scherrer studied X-ray diffraction by crystals in powder form. In this form there are always some of the minute crystals correctly orientated for Bragg reflection to occur. Thus diffraction cones are formed, with their apex at the powder (Fig. 213*a*). In practice, a thin strip of photographic film is bent into cylindrical form to give, when straightened out, a series of curved lines as shown (Fig. 213*b*). Each element forming the target material in the tube gives its own line spectrum, each line representing a given Bragg plane for a given wavelength in the element's spectrum.

The quantum theory

We have seen that both mass and charge occur only in multiples of a basic unit. Radiant energy in its various forms, from gamma rays, X-rays, ultraviolet, visible light, infrared, to radar and radio waves, is also found to be 'quantized'.

Experimental curves obtained for the emission and absorption of

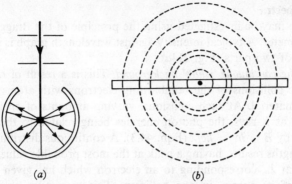

(a)   *(b)*

Fig. 213. Powder method.

radiant heat could not be explained in terms of classical wave theory. In 1900 Planck showed that, on the other hand, the curves are fitted very exactly if radiation is considered to be emitted and absorbed in small discrete packets or 'quanta' of energy, each quantum E being given by

$$E = hv \qquad\qquad\qquad\qquad\text{......................................(44.2)}$$

where v is the frequency of the radiation, and h is a constant, now called Planck's constant. Its value of 6.62×10^{-34} joule second was obtainable from the experimental curves.

The propagation of the radiation was still regarded as a continuous wave motion. But difficulty arose here also, in connection with the photoelectric effect (§ 45). In 1905, Einstein suggested that the propagation, as well as the emission and absorption, of radiation, must be regarded as quantized, in accordance with Eqn. (44.2). An independent value of h is obtainable from Einstein's photoelectric equation (p. 236) and is found to correspond exactly with that of Planck above.

It is seen that the wavelength λ of electromagnetic radiation is inversely proportional to the energy of the particle or 'photon'. Applying $v = f\lambda$ (Eqn. 30.1) to this case,

$$E = hv = \frac{hc}{\lambda} \qquad\qquad\qquad\text{.........................(44.3)}$$

where c is the velocity of electromagnetic radiation (3×10^8 m s^{-1} *in vacuo*). The energy of a photon of wavelength 10^{-10} m is thus

$$E = \frac{6.62 \times 10^{-34} \times 3 \times 10^8}{10^{-10}} = 2 \times 10^{-15} \text{ joule.}$$

MODERN PHYSICS

X-ray spectra

These may be investigated using the principle of the Bragg X-ray spectrometer. A typical intensity against wavelength graph is seen to consist of two parts (Fig. 214a).

(1) *The continuous 'white' background.* This is a result of random inelastic collisions of the bombarding electrons with atoms of the target material. At such collisions, varying amounts of energy are emitted as X-rays, the photon energies being related to the wavelengths by $E = hv = hc/\lambda$ (Eqn. 44.3). A continuous distribution of wavelengths results, having a peak at the most probable value, and a cut-off at λ_0, corresponding to an electron which has given up the whole of its energy in one collision. This maximum energy E_m associated with λ_0 clearly depends on the accelerating voltage V of the tube, and is given by $E_m = eV$ (Eqn. 43.3), where e is the electronic charge. Combining this with Eqn. (44.3) above, we obtain for the cut-off frequency v_0 and wavelength λ_0

$$E_m = eV = hv_0 = \frac{hc}{\lambda_0} \qquad \ldots\ldots\ldots\ldots(44.4)$$

For a tube voltage of 40 kV, the cut-off wavelength is thus

$$\lambda_0 = \frac{hc}{eV} = \frac{6.62 \times 10^{-34} \times 3 \times 10^8}{1.60 \times 10^{-19} \times 4 \times 10^4} = 3 \times 10^{-11} \text{ metre.}$$

(2) *The characteristic line spectra.* If the accelerating voltage is sufficiently high, some of the bombarding electrons, instead of being involved only in simple collisions, will cause ionization of the target atoms. The electrons thus ejected from the target atoms are the deep electrons from the innermost K shell, or the next L shell, etc. This is in contrast to the production of visible line spectra (p. 69) which are due to electron transitions in the outer shells of the atom, involving much less energy. As in the case of the visible spectrum, the X-rays are emitted when electrons fall back into the vacant places left by the ionization of the atom. A quantum of radiation hv is emitted as the electron falls, equal in value to the difference in energy between the two levels. Thus we obtain characteristic line spectra, as with light, which depend only on the material of the target.

Fig. 214b shows some of the energy level transitions responsible for X-ray lines. The two main peaks for all elements are the K_α and K_β lines, representing transitions to the K shell from the L and M shells respectively. As all shells except the K have a fine structure of slightly different levels, there is also a fine structure to the lines them-

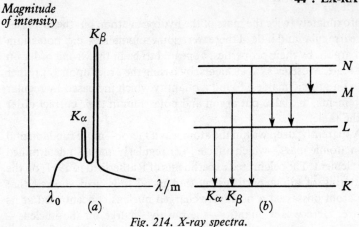

Fig. 214. X-ray spectra.

selves; for example, each K-line is a doublet. The $K_{\alpha 2}$ line results from the transition L_{II} to K. For tungsten, the measured wavelength being 2.14×10^{-11} metre for this line, by applying Eqn. (44.3) we can calculate the energy difference between these two levels for the tungsten atom,

$$E_2 - E_1 = \frac{hc}{\lambda} = \frac{6.62 \times 10^{-34} \times 3 \times 10^8}{2.14 \times 10^{-11}} = 9.3 \times 10^{-15} \text{ joule.}$$

Conclusions on atomic structure

Examination of the *optical* spectra of elements shows a periodicity as one moves up the Periodic Table, which corresponds to the periodicity of the chemical properties. This is to be expected since the outer electrons in the atom are responsible both for the optical spectra and for the chemical bonds with other atoms.

X-ray spectra are caused by transitions in the *innermost* electrons of the atom. Investigation of the K-lines in the X-ray spectra of elements shows, not a periodicity, but a progressive and regular increase in frequency v with position in the Periodic Table. Moseley in 1913 found that, for each K-line, on plotting $\sqrt{v}$ against position of the element in the table (known as its atomic number Z),

$$\sqrt{v} \propto (Z - 1),$$

i.e., $\sqrt{v}$ increases by regular steps as we pass from one element to the next.

Prior to 1913, the elements had been placed in ascending order in the table according to their relative atomic masses. The relative atomic masses so arranged increased from one to the next by approximately twice the mass of the hydrogen atom, but the increases were irregular and indeed there were some anomalies—e.g. potassium and argon, by their properties, appeared to be in the wrong order on this scale. Moseley's experiments, by basing the order upon $\sqrt{v}$ rather than on atomic masses, found a quantity which increased by regular increments, and also put argon and potassium in their correct order in the table.

Apparently; deep within the atom was a factor—more fundamental than atomic mass—which affected $\sqrt{v}$ regularly and also determined the element. The celebrated experiments of Rutherford in 1911, on the scattering of alpha particles by thin metal foils, had showed that the atom possessed a positively charged nucleus concentrated at its centre. Here was a quantity—the positive charge on the nucleus—which appeared to exist in multiples of a basic unit, and to determine the element and its position in the Periodic Table. Later and more accurate experiments by Chadwick, using the same scattering techniques of Rutherford, enabled *absolute* values of the nuclear charge to be measured. These were found to be multiples of the electronic charge.

The picture of the atom thus became clearer. Since the charge on the electrons surrounding the nucleus must equal the nuclear charge for electrical neutrality of the atom as a whole, the number of electrons was now determined also. The hydrogen ion H^+ in the discharge tube, was known to be, from measurements of its e/m' by the mass spectrograph, a hydrogen atom which had lost one electron. Since it was now established that the hydrogen atom possessed but one electron, to balance its nuclear charge, this H^+ ion, or *proton* as it was called, was clearly a hydrogen nucleus. Similarly, the same argument applied to identify the alpha particle He^{++} with the helium nucleus.

45. Photoelectric Effect

Demonstration of the effect

A zinc plate, freshly amalgamated using dilute sulphuric acid and a small drop of mercury, is connected to a gold leaf (or a pulse) electroscope, and subjected to the rays from an ultraviolet lamp.

Nothing will happen if the electroscope is positively charged, however close the lamp is placed; but if negatively charged, it will slowly discharge (or pulse), and this will be more rapid the closer the lamp to the zinc plate. The effect cannot be due to ionization of the air in the vicinity of the plate, since it is only observed if the lamp is shone directly on the plate.

The experiment shows that the ultraviolet radiation on the plate causes it to emit negatively charged particles. The identity of these particles with electrons can be established by measurement of their specific charge e/m.

Definition

The PHOTOELECTRIC EFFECT is the phenomenon of the liberation of electrons when electromagnetic radiation falls on matter.

It should be noticed that this is the converse effect to the emission of X-rays when electrons fall on a metal target (p. 225).

Quantitative investigations

In order to find what factors influence the numbers and the energies of the electrons emitted, it is necessary to be able to measure these two quantities independently.

(a) *Numbers*. As in the case of the thermionic valve (p. 210), a positive electrode collector (or anode) is placed in the vicinity of the negative plate emitting the electrons. With increasing positive potential (only a volt or two in this case) the current increases, suggesting that all the electrons have not the same velocity of emission. The main reason, however, why current increases with anode potential is that, due to the usual electrode arrangement, it is only at the higher potentials that the electrons released are increasingly guided along electric field lines which terminate at the anode.

With sufficient positive potential, saturation current is obtained, when all the electrons emitted are collected. The number of electrons emitted is proportional to this saturation current.

(b) *Energies*. If we now apply a small *negative* potential to the collector, the current is reduced, since only the faster electrons can now reach the collector against the adverse voltage. As the negative voltage is increased, a cut-off voltage occurs at which even the fastest electrons cannot reach the collector, and the current ceases. The maximum kinetic energy of emission $\frac{1}{2}mv_m^2$ is thus obtainable from Eqn. (43.3)

$$\frac{1}{2}mv_m^2 = eV_m,$$

MODERN PHYSICS

where V_m is the cut-off voltage (corrected for 'contact' potential). Thus *numbers* and *maximum energies* of the photoelectrons can be measured separately. We can now consider the effects of the *intensity* and the *wavelength* of the incident radiation on these quantities.

(1) Effect of intensity. By varying the distance of the source (inverse square law) we can vary the intensity, and it is found that:

    (*a*) the intensity is proportional to the numbers emitted;

    (*b*) the intensity has no effect on the energies.

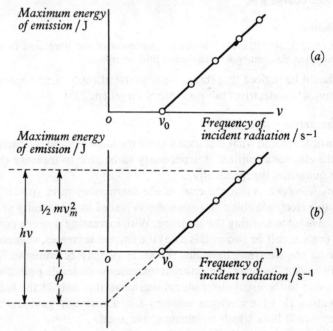

Fig. 215. Photoelectric effect. (a) Experimental graph. (b) Interpretation of $hv = \phi + 2mv_m^2$.

(2) Effect of wavelength. Using line spectra sources and filters, the wavelength (or frequency) of the source can be varied. It is found that the wavelength affects the energies of emission. A graph of maximum energy of emission against frequency of incident radiation gives the graph shown (Fig. 215a). A *threshold frequency* v_0 exists, at which the emitted maximum energy is zero, i.e. no electrons are liberated. At higher frequencies v (shorter wavelengths), the maximum energy is proportional to $(v - v_0)$.

234

Significance of these results

These results have an important impact on the wave and particle theories of light (p. 86).

(1) If the incident light were a *wave*, an increase in its intensity would imply an increase in its energy, and it would therefore eject electrons having greater individual energies. However, the energies of the electrons are unaffected (see above).

On the *particle* theory of light, an increase in its intensity implies an increase in the numbers, not the individual energies, of the incident photons. Thus, if one photon ejects one electron, the numbers, not the energies, of the ejected electrons would be affected. This is found to be the case.

It is also difficult to see how a spread-out wave could concentrate its energy sufficiently to affect a localized electron, whereas a photon could more plausibly do so.

The photoelectric effect thus sustains the particle theory of light against the wave theory.

(2) Further information is obtained from the second experiment, which yields the graph of Fig. 215*a*. This shows that the energies of the ejected electrons *can* be increased by reducing the wavelength, i.e. increasing the frequency, of the radiation source associated with the incident photons. This means that the energies of the photons themselves must depend upon the wavelength or frequency associated with them.

The straight line graph (Fig. 215*a*) can be interpreted even more precisely, i.e. in terms of $E = hv$ (Eqn. 44.2), which asserts that the energy of a photon is not merely dependent on the frequency of the radiation, but *proportional* to it. If we continue the graph backwards we obtain Fig. 215*b*. The total energy E of the incident photon we will now call, according to Eqn. (44.2),

$$E = hv,$$

where h is Planck's constant. Some of this energy is used to remove the electron from the surface of the metal. This is called the 'work function' ϕ.

The WORK FUNCTION (ϕ) of a metal is the least additional energy required to enable an electron to escape from the surface.

The balance of the energy appears as kinetic energy of emission. If a photon interacts directly with a free electron on the surface, the electron will escape with the *maximum* kinetic energy $\frac{1}{2}mv_m^2$.

Clearly, therefore,

$$hv = \phi + \frac{1}{2}mv_m^2$$

235

At the threshold frequency v_0, the maximum kinetic energy is zero, so

$$hv_0 = \phi.$$

Combining these two equations, we have Einstein's photoelectric equation,

$$\tfrac{1}{2}mv_m{}^2 = h(v - v_0) \dots\dots\dots\dots\dots(45.1)$$

which is the equation of the graph Fig. 215a.

It is seen that the photoelectric effect thus fits in convincingly with the quantum equation $E = hv$. An independent value of Planck's constant h can be obtained from the slope of the experimental graph.

46. Radioactivity

Discovery

In 1896, Henri Becquerel found that crystals of a certain uranium salt caused blackening of a photographic plate placed in their vicinity. Apparently spontaneous radiations were being emitted from the crystals—radiations that needed no external source of energy. After several months the activity was still present without measurable reduction.

Like X-rays, which had been discovered only the year previously, these radiations could pass through layers of opaque materials, and could ionize gases. Further investigations showed that the significant component of the source was the element uranium, and that all thorium salts likewise produced these effects. A few years later, Pierre and Marie Curie isolated from large quantities of pitchblende small amounts of two new and more active elements, which they named polonium and radium. Later, a third new and active element, named actinium, was isolated from pitchblende residues. These three new elements all filled gaps in the Periodic Table. The phenomenon was given the name of 'radioactivity'.

PROPERTIES OF ALPHA, BETA, AND GAMMA RADIATIONS

Nature of the radiations

Alpha particles are helium nuclei, consisting of two protons plus two neutrons, with a mass of 4 units and a charge of $+2$ units. Beta particles are fast-moving electrons, having a rest mass of $\frac{1}{1836}$ unit and a charge of -1 unit. At high velocities, approaching that of light, the mass increases according to relative theory.

Gamma rays are high-energy electromagnetic waves, possessing neither mass nor charge. The wavelength of gamma rays is shorter than that of X-rays, ranging from about 4×10^{-10} to 5×10^{-13} m.

Properties by which detected

The radiations possess three basic properties by which they may be detected:

(1) They affect photographic plates. As well as a general blackening of photographic film, individual tracks of alpha and beta particles can be recorded. Owing to the high absorption of the emulsion, these tracks are not more than a few thousandth of a cm in length.

(2) They cause scintillations in substances such as zinc sulphide. Individual alpha particles are observed as tiny flashes of light. Prior to 1930, this tedious counting method was virtually the only way of studying alpha particles quantitatively. The *spinthariscope* is a device for observing these scintillations, which well illustrate the randomness of the emission. A speck of radium is placed a few millimetre from a zinc sulphide screen at one end of a tube, a viewing lens being at the other. The eye needs adjustment to the dark before viewing.

(3) They ionize gases, This property gives rise to a number of detecting instruments (pp. 238–42).

Mechanism of ion-pair production in gases

Both alpha and beta particles eject electrons from gaseous molecules in their paths, forming ion-pairs consisting of an electron and a relatively heavy positive ion. In certain circumstances, the liberated electron may attach itself to a neutral molecule forming a negative ion. In this case the positive and negative ions are of comparable mass and mobility.

Gamma rays ionize gases more indirectly. The electrons ejected initially by the rays are fast-moving. These, in turn, eject further electrons from molecules along their paths, and this secondary effect is the source of most of the ion-pairs produced.

The effect of ionization on range. For each ion-pair produced, a particle expends an approximately constant amount of its kinetic energy. Alphas and betas having similar initial energies of emission will therefore produce roughly the same total number of ion-pair before being brought to rest. But since an alpha is some 8000 times as heavy as a beta, its velocity is less, and it produces more ion-pair per metre of its path. Consequently, alphas produce a far heavier ionization over a much shorter range.

In fact, alphas from radioactive sources produce about 10^7 ion-pair per metre in ordinary air, and have a range of only a few centimetre. They are stopped entirely by a thin metal foil or a few sheets of paper, or by the skin.

Betas of similar energies produce only a few thousand ion-pair per metre, and have ranges several hundred times those of alphas. They are stopped by a few millimetre of aluminium.

Gamma rays ionize still less, and are therefore very penetrating. Several centimetre of metal are necessary to reduce the intensity to undetectable amounts.

Ionization current. Normally the ion-pairs produced soon recombine, but if the ionized gas is situated between two electrodes across which a high p.d. is applied, the positive and negative ions move towards opposite electrodes and an ionization current results. This is extremely small, about 10^{-10} A in school experiments, but is measurable. Such a current indicates the presence of an ionizing agency, which may be a radioactive source, a beam of X-rays or ultra-violet rays, or even a flame.

Detecting instruments

The following instruments detect radiations by the ionization produced.

The gold leaf electroscope. If the electroscope is charged in the normal way, and a radioactive source brought near, those ions formed in the air that are of opposite charge to the electroscope are attracted towards it, and the leaf gradually falls as the electroscope is discharged.

The quartz fibre electroscope. A fine metal-coated quartz fibre replaces the gold leaf of the previous instrument. The insulated part of the system, which includes the quartz fibre, is initially charged with respect to the metal case by connecting momentarily across an H.T. battery. The fibre is repelled, as in the case of the gold leaf, and upon exposure to ionizing radiations it gradually returns to its former position. The deflection is observed through a lens system. The total amount of movement is a measure of the cumulative radiation received since the last charge, and the instrument, like the photographic film, can be carried around as a personal monitor.

The Wulf, or pulse, electroscope. This is in principle a gold leaf electroscope with a counter-electrode C added (Fig. 216a). The 'leaf' L (shown diagrammatically) is some form of flexible component which can be attracted towards the counter-electrode. On touching C, the 'leaf' is re-charged. Continuous pulsing can occur, indicating an ionization current of the order of microampere.

There are three electrodes. The outer case E and electrode A are earthed. The counter-electrode C is maintained at a high potential, usually $+3$ to $5\,\text{kV}$. The central insulated electrode B, with the leaf L, is at a floating potential somewhere between that of A and C. When

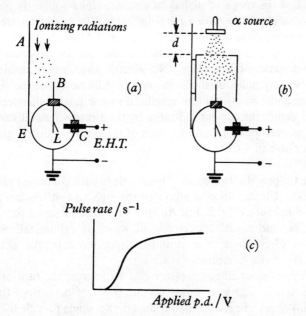

Fig. 216. Pulse electroscope (a) without and (b) with ionization chamber.

C is connected to the E.H.T. a single pulse occurs, causing B to attain the potential of C. A high p.d. now exists between A and B, and when radiations cause ionization of the air between them, the ionization current across AB causes the charge in B to leak away. This in turn results in a p.d. between B and C, the leaf is again attracted towards C, a pulse takes place, and the cycle is repeated. The rate of pulsing is clearly proportional to the ionization current.

For quantitative work, an *ionization chamber* is fitted to A (Fig. 216b). The outer case thus becomes the earthed electrode, with B as the central positive electrode. If radiations, from an alpha source placed above it, are allowed to enter the chamber through a wire mesh, a characteristic curve of ionization current (or pulse rate) against applied p.d. can be drawn. The position of the alpha source is kept constant immediately above the wire mesh (i.e. $d = 0$). As the applied p.d. is increased the pulse rate increases, until it attains a fairly steady value independent of the p.d. (Fig. 216c). This clearly represents the situation in which all ions produced are collected at the electrodes before appreciable re-combination can take place. The ionization chamber is always used on this 'plateau'.

To find the range of alphas in air. An alpha source is placed a distance d above the wire mesh (Fig. 216b). The side electrode of the pulse electroscope is maintained at about 3 kV, sufficient to ensure that all the ions produced within the chamber are collected before re-combination takes place. No electric field exists outside the chamber, so none of the ions outside will be collected. As d is increased, the number of ions produced within the chamber decreases, consequently the current indicated by the electroscope falls off. The point of zero pulse rate on a graph of pulse rate against d gives the alpha range in air.

The Geiger-Müller counter. This is a device for counting individual particles. The radiation is admitted through an end window E into a gas-filled tube (Fig. 217a). An H.T. source maintains a p.d. across the tube, the negative electrode of which is cylindrical, and the positive electrode a wire running along the axis. An electronic counting device completes the circuit.

The conditions of gas pressure and p.d. across the tube are such that a spark discharge is just on the point of occurring. In these circumstances, the ionization produced by a single particle is enough to trigger off an avalanche of electrons (by secondary ionization)

240

which will cause a sizeable pulse of electricity to pass round the circuit.

The filling gas is a mixture of argon, for initiating the discharge, and a small quantity of another gas, for quenching the discharge quickly. Tubes quenched by organic gases operate at voltages up to 1500 V, but halogen quenched tubes can operate at about 400 V. Even with rapid quenching in a time of the order of 10^{-4} s, the arrival of two particles almost together may result in only a single pulse being registered.

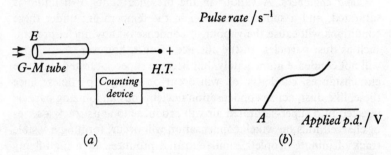

Fig. 217. The Geiger-Müller counter.

The p.d. across the tube is adjusted to the 'plateau' region (Fig. 217*b*), so that variations in p.d. will not affect the count rate. Below a certain p.d. (point *A*) no pulses are registered, since the counting device in the circuit is designed to register only pulses above a certain magnitude. As the p.d. is increased, the pulse rate for a given source increases since the secondary ionization by electrons increases. A plateau is then reached on which the pulse rate is approximately independent of the p.d. applied. A further increase in p.d. would result in a continuous spark discharge across the tube.

The counting device may be one of two types. A *scalar* circuit contains units each of which divides the number of incoming pulses by (say) 10, thus reducing them to a number capable of being registered on a mechanical counter. The total number of pulses is recorded over a timed period, after which the counter can be re-set to zero for the next count. A *ratemeter* indicates the average count rate on a galvanometer scale. Various ranges can be set. A time-constant adjustment allows the integration of counts to take place over various periods of time. For example, if the reading is allowed to build up over only a few second, it may then be erratic due to random variations in the count rate; if a larger time-constant is used, a more stable reading is obtained, but with a longer wait for the reading to build up.

The G.M. tube is unable to distinguish between the various ionizing particles. It is usually used for beta and gamma rays. A specially thin window is needed for the tube for alpha particles, since these are easily absorbed.

A loudspeaker can usually be switched into the circuit so that individual clicks can be heard, as well as recorded electronically.

Cloud chambers. A vapour in the presence of its own liquid is saturated, and a sudden reduction in the temperature under these conditions will cause the vapour to condense on any nuclei present, such as dust particles. In the absence of dust, however, the vapour will not condense immediately, but become super-saturated. In these circumstances, condensation will occur on charged particles, since these, like dust, act as condensation nuclei. Thus an ionizing particle injected into super-saturated air will produce along its track a series of charged ions on which condensation will occur, forming a visible track of minute droplets, similar to that produced by a high-flying aircraft.

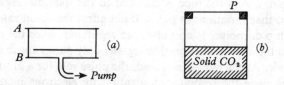

Fig. 218. Cloud chambers. (a) Wilson (b) Continuous.

In the *Wilson cloud chamber* (Fig. 218a), a few cubic centimetre of alcohol produce a saturated vapour, a sudden expansion of which is effected with a pump, e.g. a reversed bicycle pump. This adiabatic expansion is accompanied by a reduction in temperature, and consequently super-saturation occurs. Tracks from an alpha particle source placed inside the chamber are clearly seen momentarily, when viewed from above with illumination from the side. The expansion can be repeated at intervals of a few second. An electric field must first be applied across the chamber (about 400 V between A and B) in order to remove stray ions. These would otherwise cause a cloud when the expansion was applied, and mask the tracks.

In the *continuous cloud chamber* (Fig. 218b), alcohol is introduced into a ring of absorbent material at the top of the chamber. 'Dry ice' (solid CO_2) is placed immediately beneath the chamber, setting up a vertical temperature gradient. The vapour, initially saturated, falls into regions of lower temperature and becomes super-saturated. Tracks can be produced in this region. Unlike in the Wilson type chamber, tracks are continually visible. The initial clearing of ions by an electric field is effected, in a simple version of this apparatus, simply by rubbing the upper plastic plate P.

The absorption of alpha, beta and gamma radiations

We have seen (p. 238) that alpha particles produce heavy ionization and have the shortest range, while gamma rays ionize least and are therefore the most penetrating. The various factors involved when the three radiations encounter matter—whether gaseous or a solid absorber—are now considered in a little more detail.

Alpha absorption. Cloud chamber tracks of alpha particles show a uniform range, indicating a uniform energy of emission from the radioactive source. The particles also show little scattering, as is seen by their undeviated paths. The ionization occurs more heavily towards the end of the track, where the particle is slowing down; the ionization then rapidly diminishes to zero, giving a well defined end-point. The interposing of an absorber, such as a sheet of paper, slows down *all* the particles, reducing their range. It does not reduce appreciably the number of particles that will arrive at the end-point.

Beta absorption. Unlike alphas, beta particles are emitted with a continuous distribution of velocities up to a definite maximum value. Also unlike alphas, the scattering of betas by matter is very marked. Thirdly, in addition to the loss of energy by ionization, betas may also collide with molecules without causing ionization, but will suffer a deceleration, and the balance of energy is emitted in this case in the form of electromagnetic radiation. (Compare the production of X-rays, p. 225). This is known as the *Bremstrahlung effect*.

With all these factors, it is seen that the absorption of beta particles is a complicated process.

Gamma absorption. Gamma rays are not absorbed appreciably by the air, and investigation of the variation of intensity with distance from a point-source in air, which can be carried out with a Geiger counter, reveals an inverse square law, as with light (p. 72). Beta rays, on the other hand, owing to their absorption and scattering by the air, do not obey this law.

The absorption of gamma rays by matter involves a number of processes. Firstly, by the photoelectric effect (p. 233) the gamma ray photon loses the whole of its energy in the ejection of an electron from an atom or molecule of the absorbing material. In a second process, called the *Compton effect*, the photon collides with an electron, is deflected, and loses part of its energy to the electron which recoils as if struck by a Newtonian particle. A third process is the formation of positron-electron pairs, resulting in the annihilation of the incident photon. Pair production occurs only if the energy of the gamma radiation exceeds 1 MeV, or 1.6×10^{-13} joule.

In all these processes fast electrons are produced. These in turn, as was seen on p. 237, cause the ionization by which the gamma rays are detected.

Mathematical treatment of absorption

The fractional reduction in intensity produced by an absorber will depend upon: (*a*) the type of radiation, (*b*) its energy or wavelength, (*c*) the material of the absorber, and (*d*) its thickness. For a given radiation and absorbing material we can define an absorption coefficient as follows:

The LINEAR ABSORPTION COEFFICIENT (μ) is the fractional reduction in intensity, per unit thickness of absorber. *Unit:* per metre.

Expressed mathematically,

$$\mu = -\frac{\delta I}{I \cdot \delta s},$$

where I is the observed intensity for a thickness s, and $-\delta I$ is the decrease in intensity due to an additional thickness δs.

Separating variables and integrating between the limits shown, where I_0 is the intensity for $s = 0$,

$$-\mu \int_0^s ds = \int_{I_0}^I \frac{dI}{I},$$

giving
$$-\mu s = \log_e I - \log_e I_0 \quad \dots\dots\dots\dots\dots(46.1)$$

which becomes
$$I = I_0 e^{-\mu s} \dots\dots\dots\dots\dots\dots\dots(46.2)$$

i.e., an exponential decay curve (Fig. 219*a*).

Converting to $\log_{10}$ and re-arranging,

$$\log_{10} I = -0.434 \mu s + \log_{10} I_0 \dots\dots\dots\dots(46.3)$$

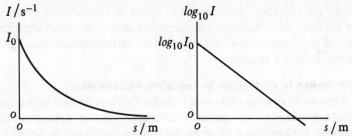

Fig. 219. (a) Exponential decay curve. (b) Straight line logarithmic graph.

which is a straight line graph of the form $y = mx + c$ (Fig. 219b). The value of the linear absorption coefficient is obtainable from the slope.

Application of the logarithmic absorption law. The above theory assumes the absorption coefficient μ to be constant, i.e. independent of the thickness s of the absorber. The question arises: to what extent is this true for alpha, beta, and gamma rays?

For gamma rays, the theory applies strictly, and can even be extended. Dense materials, such as lead, are more effective absorbers of gamma rays than light materials. In fact, the effect is approximately proportional to the density of the absorber. We may therefore define a further quantity which, for low-energy gamma rays at least, is almost independent of the material of the absorber:

The MASS ABSORPTION COEFFICIENT is the linear absorption coefficient divided by the density of the absorber. *Unit:* $m^2\,kg^{-1}$.

Beta particles, in view of the complications already described (p. 243), could hardly be expected to obey a logarithmic law of absorption. By a combination of circumstances, however, they do so approximately. It is therefore possible to calculate linear absorption coefficients for these particles also. The absorption, again, depends on the density of the absorber, and a mass absorption coefficient may be found which under restricted conditions is approximately independent of the absorber.

In the case of alpha particles, an absorber does not reduce the number of particles, only their speed (p. 243). A detector that counted individual particles, for example a Geiger counter, would therefore certainly not indicate a logarithmic law of absorption. The ionization chamber connected to the pulse electroscope does not, however, count particles, but records the total amount of ionization occurring within the chamber. A reduction in speed of the particles entering the chamber thus results in a reduced ionization current. A logarithmic plot of alpha absorption in these circumstances does give approximately a straight line.

Experiments to distinguish the radiations by absorption

A source of all three radiations (e.g. Ra 226) may be used with (*a*) an ionization chamber and pulse electroscope to detect alpha and beta particles (the ionization of air by the gamma rays being negligible by comparison), and with (*b*) a Geiger tube with a rate-meter or scaler counter to detect beta and gamma rays (in this case it is the alphas which are undetected since they are absorbed by the end-window of

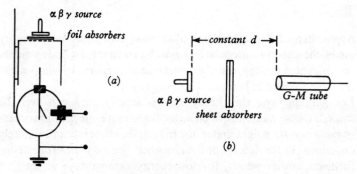

Fig. 220. *Absorption experiments.*

the tube). Absorbing screens used in the two cases reveal in each case two distinct radiations, the absorption coefficients of which can be calculated (p. 245). The absorption coefficient of the *more* penetrating radiation in (*a*) is found to be the same as that of the *less* penetrating radiation in (*b*).

These two experiments thus suggest that there are, in fact, these three distinct radiations (which we call alpha, beta, and gamma) from Ra 226.

(*a*) **To distinguish alpha and beta radiations.** The source is placed close to the wire mesh of the ionization chamber connected to the pulse electroscope (Fig. 220*a*). Increasing thicknesses *s* of a thin aluminimum foil are inserted between the source and mesh, and the pulse rate *I* noted in each case. A graph of $\log_{10} I$ against *s* is plotted, and is found to be of the form shown in Fig. 221, showing that two different radiations are involved. The two absorption coefficients are found as described below.

(b) **To distinguish beta and gamma radiations.** The same source as above is used. In this case, the detector is a Geiger tube (Fig. 220b). Increasing thicknesses of aluminium, up to several millimetre, are interposed, and the count rate I noted in each case. As before, a graph of $\log_{10}I$ against thickness s is plotted, and is again found to be of the form shown in Fig. 221. The absorption coefficients are found in the same way as before.

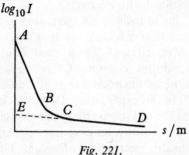

Fig. 221.

Calculation of absorption coefficients. If I' and I'' are the separate intensities contributed by the two radiations, for various thicknesses s of absorber, then we may write (Eqn. 46.3)

$$\log_{10}I' = -0.434\mu's + \log_{10}I'_0 \qquad (46.4)$$

and
$$\log_{10}I'' = -0.434\mu''s + \log_{10}I''_0 \qquad (46.5)$$

But the total *observed* intensity I is given by

$$I = I' + I'' \qquad (46.6)$$

for each value of s.

A graph of $\log_{10}I$ is plotted against s (Fig. 221). The portion AB is due to both radiations, but CD is due to the more penetrating radiation only. Thus CD is expressed by Eqn. (46.5), from the slope of which the absorption coefficient μ'', of the more penetrating radiation, can be found.

The portion CD is now produced back to E, and for various values of s between E and C $\log_{10}I''$ is read off, and I'' deduced in each case. The corresponding values of I are known from AB, and hence I' values can be calculated from Eqn. (46.6).

This enables a plot of $\log_{10}I'$ against s to be made over this region. Hence the absorption coefficient μ', of the less penetrating radiation, can be deduced from the slope, by Eqn. (46.4).

MODERN PHYSICS

Identification of alpha, beta, and gamma rays

The above absorption experiments indicate the presence of three distinct types of radiation. Other evidence is needed to show the exact nature of these radiations.

Beta particles are found to be readily deflected by electric and magnetic fields, in such a way as to indicate they carry a negative charge. In these experiments, their velocities and value of e/m can be measured. Other evidence confirms that, like cathode rays, beta particles can be identified with electrons.

Alpha particles, due to their larger mass, are far more difficult to deflect, but the direction of deflection in strong magnetic fields shows that they are positively charged. Again, their value of e/m can be measured. The conclusive experiment, however, identifying the alpha particle with the helium nucleus was carried out by Rutherford and Royds in 1909. In this experiment, alpha particles from a radioactive material were allowed to pass through a thin glass tube into an evacuated space. Subsequently, an electrical discharge in this space showed the unmistakable spectrum of helium gas. The only source of this helium could be the alpha particles.

Gamma rays are not deflected by magnetic or electric fields. Definite proof of their nature was obtained in 1914, when the wavelengths of the rays were measured by the crystal diffraction method (p. 227). They were found to be, like X-rays, electromagnetic waves of very short wavelength.

PROPERTIES OF RADIOACTIVE SUBSTANCES

Structure of the atom

Evidence for the basic structure of the atom comes from many sources and cannot be summarized here. The evidence shows that atoms are made up of a relatively massive nucleus consisting of protons (mass 1 unit, charge +1 unit) and neutrons (mass 1 unit, charge zero) surrounded by orbital electrons (of negligible mass, charge −1 unit).

The ATOMIC NUMBER (Z) of an atom is the number of protons in the nucleus, i.e. its number of *charge* units. This is equal to the number of orbital electrons when the atom is in its neutral state.

The atomic or charge number Z determines the chemical properties of the atom, and the element to which it belongs in the Periodic Table. For example, hydrogen has $Z = 1$, helium $Z = 2$.

The MASS NUMBER (A) of an atom is the number of protons plus the number of neutrons in the nucleus, i.e. its number of *mass* units.

Atoms of light elements are made up of an approximately equal number of protons and neutrons; for heavy elements the ratio of neutrons to protons approaches 1.5. For a given element, however the possible number of neutrons in the atom is variable, giving rise to a number of different *isotopes* of that element.

ISOTOPES of a given element possess the same number of protons, but a different number of neutrons.

For example, carbon has six isotopes altogether, including:

Carbon 11	6 protons + 5 neutrons	$Z = 6$	$A = 11$
Carbon 12	6 protons + 6 neutrons	$Z = 6$	$A = 12$
Carbon 14	6 protons + 8 neutrons	$Z = 6$	$A = 14$

Note that:

(1) All the isotopes of a given element are chemically identical and therefore chemically indistinguishable. This is because the chemical properties depend upon the number of orbital electrons, which is equal to the number of protons in the nucleus, i.e. is the same for all isotopes of the same element.

(2) Some isotopes occur naturally, while others may be produced artificially, e.g. by neutron bombardment of natural isotopes.

(3) Some isotopes are stable (e.g. carbon 12), while others are unstable (e.g. carbon 11 and 14). It is the unstable isotopes that are 'radioactive'.

Natural radioactive series. An atom of an unstable isotope ejects, sooner or later, either an alpha or a beta particle, together perhaps with some gamma rays. The atom is thereby transmuted into a different element, since the Z number is changed. The new isotope may itself be unstable, and several further radioactive emissions may occur before final stability is attained.

There are four naturally occurring radioactive series, called the uranium, thorium, actinium, and neptunium series. These are all in the higher reaches of the Periodic Table, since atoms of high Z numbers are more prone to instability. The end-product of the neptunium series is a stable isotope of bismuth. Each of the other three terminates in a stable isotope of lead. The thorium series is given for illustration:

Table IV—Thorium Series

Isotope	Z number	A number	Particle emitted	Half-life
Thorium 232	90	232	alpha	1.4×10^{10} year
Radium 228	88	228	beta	6.7 year
Actinium 228	89	228	beta	6.13 hour
Thorium 228	90	228	alpha	1.91 year
Radium 224	88	224	alpha	3.64 day
Radon 220	86	220	alpha	52 second
Polonium 216	84	216	alpha	0.16 second
Lead 212	82	212	beta	10.6 hour
Bismuth 212	83	212	beta* and alpha†	60.5 minute
*Polonium 212	84	212	alpha	3×10^{-7} second
†Thallium 208	81	208	beta	3.1 minute
Lead 208	82	208	stable	......

Decay constant and half-life

It is observed in experiments with the spinthariscope, Geiger counter, etc., that the emission of individual alpha and beta particles is a random process, the instant of decay of any particular atom being unpredictable. Averaged over a large number of atoms, however, the proportion likely to decay in a given period of time is predictable, some isotopes being far more quick to decay than others. The measure of the readiness of a particular isotope to decay is expressed in terms of the decay constant λ and the half-life $T_{\frac{1}{2}}$.

The DECAY CONSTANT (λ) of a radioactive isotope is the fraction of its atoms decaying per unit time. *Unit:* per second.

This quantity is found to be independent of all conditions. It is, therefore, a sub-atomic phenomenon, since all atomic phenomena depend on such conditions as temperature, pressure, etc.

Expressed mathematically,

$$\lambda = -\frac{\delta N}{N \cdot \delta t} \quad \text{.........................(46.7)}$$

where N is the number of atoms of the isotope present at the time t, and $-\delta N$ is the decrease in number due to decay in the time δt.

Separating variables and integrating,

$$-\lambda t = \log_e N - \log_e N_0 \text{.....................(46.8)}$$

which becomes $\qquad N = N_0 e^{-\lambda t} \quad \text{..........................(46.9)}$

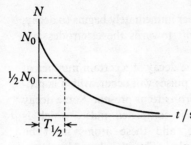

Fig. 222. Exponential decay and half-life $T_{\frac{1}{2}}$.

where N_0 is the number of atoms of the isotope present at the time $t = 0$. The analogy with Eqn. (46.2) is evident.

The exponential decay curve (Fig. 222) is asymptotic to the t axis and clearly there is no useful meaning to the term the 'life' of a radioactive isotope. The 'half-life', however, is meaningful:

The HALF-LIFE ($T_{\frac{1}{2}}$) of a radioactive isotope is the time taken for the number of its atoms to decay to one half of its initial value.

Converting Eqn. (46.8) to $\log_{10}$, and re-arranging,

$$-0.434\lambda t = \log_{10}\frac{N}{N_0},$$

and from the definition of half-life $T_{\frac{1}{2}}$,

$$-0.434\lambda T_{\frac{1}{2}} = \log_{10}\tfrac{1}{2},$$

giving $\qquad \lambda T_{\frac{1}{2}} = 0.693$ ....(46.10)

To find the half-life of thoron. The squeeze bottle S (Fig. 223a) contains thorium hydroxide. Squeezing the bottle releases radioactive thoron gas into the ionization chamber. The gas is Radon 220 in the Thorium Series (p. 250), having a half-life of about 52 second. The radioactive equilibrium (p. 252) in the bottle will subsequently take about a week to be restored by the decay of Radium 224, which has a half-life of 3.64 day.

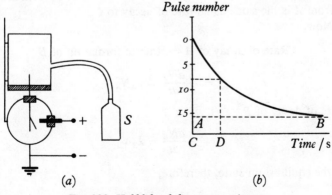

Fig. 223. Half-life of thoron experiment.

The thoron in the ionization chamber immediately begins to decay, producing ions in the air, which drift towards the electrodes and produce pulsing of the electroscope.

Each completed pulse represents the decay of a certain number of thoron atoms, and a total number of pulses will occur, which is proportional to the initial number of thoron atoms present. Some decay will continue, however, after 'the last completed pulse, which is insufficient to cause another pulse, and these atoms must be allowed for by a final 'fraction of a pulse'. This is deduced by drawing the asymptote AB (Fig. 223b).

A graph of pulse number against the time at which it occurs is drawn. The rate of decay decreases proportionally to the number of active atoms still present, i.e.,

$$\frac{dN}{dt} \propto N \text{ (Eqn. 46.7)}.$$

Therefore, the rate of pulsing gradually decreases exponentially with time.

The half-life CD is the time taken to reach half the total number (including the final fraction) of pulses obtained.

Radioactive equilibrium. In a radioactive series, the radioelement A decays to B, and B decays to C, and so on, these events taking place simultaneously.

Consider first a series consisting only of A, B, and C, where initially only A is present, and this is so long-lived that the amount present remains sensibly constant. Eventually, an equilibrium state will be reached for B. This will occur when the rate of formation of B from A is the same as its rate of decay to C.

Now,

$$\text{Rate of decay of } A = \text{Rate of formation of } B$$

$$= -\frac{dN_A}{dt} = \lambda_A N_A.$$

Rate of decay of B

$$= -\frac{dN_B}{dt} = \lambda_B N_B.$$

In the equilibrium state, therefore,

$$\lambda_A N_A = \lambda_B N_B.$$

The effect of the decay of further daughter products will clearly produce the relationship, in the equilibrium state,

$$\lambda_A N_A = \lambda_B N_B = \lambda_C N_C = \text{etc} \ldots$$

In nature, the proportions of atoms (N) of the various products present in a given sample are thus related to the decay constants (λ) and the half-lives ($T_{\frac{1}{2}}$) thus, e.g.

$$\frac{N_A}{N_B} = \frac{\lambda_B}{\lambda_A} = \frac{T_{\frac{1}{2}A}}{T_{\frac{1}{2}B}}.$$

Strength of a radioactive source

The strength of a radioactive source depends upon its mass, and how radioactive it is.

The CURIE is the quantity of any radioactive material giving 3.70×10^{10} disintegration per second.

This is based on the strength of 1 g of pure radium. The strength of sources used in schools is of the order of 5 microcurie.

Dose of radiation

The dose of radiation received is a cumulative effect, and has been measured in various units. The original unit was the röntgen:

The RÖNTGEN is the quantity of X-rays or gamma rays which will produce, by ionization, 2.08×10^9 ion-pair in 1 cm³ of dry air at s.t.p.

The röntgen is equivalent to the absorption of 8.6×10^{-3} joule of energy by 1 kg of air. In soft body tissue, the absorption is about 9.7×10^{-3} J kg^{-1}.

An additional complication is that different radiations do not have equivalent biological effects. Consequently, a unit called the rem ('röntgen equivalent man') has been defined to equate the biological effects of the various radiations:

The REM is the quantity of any given radiation which produces the same biological damage in man as that resulting from the absorption of 1 röntgen of X-rays or gamma rays.

Permissible radiation dosage. We are continually subject to cosmic rays and natural radioactivity in materials around us, such as luminous paint. It is evident that restricted exposure to radiation is not dangerous, though even the smallest radiations may possibly have some genetic effects. The tolerance dose is about 0.3 röntgen per week, applying to the whole body, both externally and internally. The absorption of about 500 röntgen is a fatal dose. In addition to external radiations (beta, gamma, and neutrons), there is the danger of ingestion by inhalation, food, or through the skin. Different radionuclides also have greater or lesser danger, depending upon whether they are naturally eliminated from the body or tend to concentrate in certain organs.

Safety measures in the school laboratory. The degree of precaution needed depends upon the strengths of radiation likely to be encountered. The precautions needed for work with curie-strength sources are vastly different from those with microcurie sources. However, certain special disciplines must be observed in all cases. One of the sinister features of radioactivity is that exposure to it is not normally felt at the time: the defence mechanism of pain does not operate.

In the school laboratory, sources must at all times be handled only with special long forceps, and must be kept at a distance of at least one foot from the person. They must be returned after use to a special lead container (which seals the source in case of fire) and stored in a locked and marked cupboard. Additional precautions are needed if unsealed (or liquid) sources are used.

Index to Laws and Definitions

255

INDEX

256

INDEX